Thermal and Fluids Systems Six-Minute Problems

Third Edition

Daniel C. Deckler, PhD, PE

PPI®
PPI2PASS.COM
A KAPLAN COMPANY

Register Your Book at ppi2pass.com

- Receive the latest exam news.
- Obtain exclusive exam tips and strategies.
- Receive special discounts.

Report Errors for This Book

PPI is grateful to every reader who notifies us of a possible error. Your feedback allows us to improve the quality and accuracy of our products. Report errata at **ppi2pass.com**.

THERMAL AND FLUIDS SYSTEMS SIX-MINUTE PROBLEMS
Third Edition

Current release of this edition: 1

Release History

date	edition number	revision number	update
Jul 2017	1	1	Title change. New cover. Minor corrections.
Apr 2018	2	1	New edition. Content update. Copyright update.
Apr 2019	3	1	New edition.

Printed in the United States of America.

PPI
1250 Fifth Avenue, Belmont, CA 94002
(650) 593-9119
ppi2pass.com

ISBN: 978-1-59126-650-1

F E D C B A

Table of Contents

About the Author

Daniel C. Deckler was born in Tiffin, Ohio, and raised in various parts of Ohio. He received his bachelor's degree in mechanical engineering from Ohio Northern University, and received both his master's degree in mechanical engineering and his PhD in engineering from the University of Akron. He received his professional engineer's license in 1999.

Dr. Deckler is currently a professor of engineering at the University of Akron Wayne College, where he teaches courses in mechanical, electrical, and civil engineering. Before coming to the University of Akron, he was an engineer for Rockwell International in Downey, California, the Timken Company in Canton, Ohio, and the Loral Corporation (formerly Goodyear Aerospace) in Akron, Ohio. With these companies he worked on interesting projects such as the space shuttle and space station, bearing applications, and anti-submarine warfare systems, respectively.

Dr. Deckler lives in Canton, Ohio. When not working, he can be found enjoying the outdoors, be it golfing, cycling, running or training for triathlons, or hunting.

Preface and Acknowledgments

The Principles and Practice of Engineering examination (or PE exam for short) is prepared by the National Council of Examiners for Engineering and Surveying (NCEES). The PE exam tests examinees' understanding of both theoretical and practical engineering concepts. For mechanical engineering, three versions of the exam are given, one for each of the disciplines of HVAC and refrigeration, machine design and materials, and thermal and fluids systems. This book is designed to help you with the thermal and fluids systems exam.

For each exam, the NCEES issues a list of the knowledge areas that will be covered. The topics covered in *Thermal and Fluids Systems Six-Minute Problems* are the same knowledge areas as in the NCEES specifications for the thermal and fluids systems exam. The problems in this book cover fluid mechanics, heat transfer, mass balances, thermodynamic cycles, hydraulic and fluid equipment, energy and power equipment, cooling and heating, and more.

The third edition includes 18 problems to address the most recent code updates. I have also reorganized the problems so that the order of topics follows the order of knowledge areas in the current NCEES specifications, which became effective in April 2017. I hope this will help you focus your study to the needs of the current exam.

Some of the problems in this book originated in my professional practice, and others in my experience as an instructor. This book serves as a companion to the *Mechanical Engineering Reference Manual* and is designed primarily to help you remember what you have already learned. I have purposefully left "shortcuts" out of the problem solutions. Instead, I have focused on solving the problems from fundamental concepts. Solving in this manner might take a few more minutes, but it eliminates the need to remember special cases. One only has to recall a few basic ideas that are easily remembered or located. In addition, for several of the problems I have shown multiple methods of solution, to further strengthen your review.

The reason for this book is to give you practice in solving exam-like problems within exam-like time limits. The problems are designed to be solvable in an average of six minutes each, which is the average amount of time you'll have for each problem on the actual exam. Some will take you less than that, of course, and others will take you longer. But overall, you should be striving for an average time of six minutes or less per problem.

The problems presented in this book are representative of the type and difficulty of those you will encounter on the PE exam. The book's problems are both conceptual and practical, and they are written to provide varying levels of difficulty. Though you probably won't encounter exam problems on exactly the same situations, reviewing these problems and solutions will increase your familiarity with the exam problems' form, content, and solution methods. This preparation will help you considerably during the exam.

Problems and solutions have been carefully prepared and reviewed to ensure that they are appropriate, understandable, and correctly solved. If you find errors or discover alternative, more efficient methods of solving a problem, please bring it to PPI's attention so that your suggestions can be incorporated into future editions. You can report errors and keep up with the changes made to this book, as well as changes to the exam, by going to PPI's website at **ppi2pass.com**.

This book would not have been possible without the contributions of the many people who supported me through several years of writing and review.

A huge thank you goes to Jan Deckler. She is an engineer at a refinery, and many of the problems you see in this book are the direct result of her experience. So, whenever you see a problem about a refinery, you will now know where it came from. Another thank you goes to all my former engineering coworkers who were kind enough to take me "under their wings" and teach me what they knew when I was a beginning engineer. Believe it or not, they were an inspiration for several problems so many years later. Finally, I would like to thank my former and current students who are interested in real-life applications of the subjects I teach. They are always making me reach into my memory and pull out useful problems that I had to solve during my days in industry and as a consultant. Several of these problems have found their way into this book.

From PPI, I recognize Aline Magee, who started me along this journey; Sarah Hubbard, who helped me along the way; and Scott Marley, who persisted with me to the completion of this book. Scott is a fiend about deadlines, and there were times when I found him infuriating. Scott, however, has the patience of a saint and stuck with me despite my failings, and for that I owe him a great deal of thanks. Thank you, Scott, this book would never have been completed without your dedication. To David Bostain, Matt Gordon, and Muamba

Wanzala, who reviewed and helped polish the manuscript, Anil Acharya, calculation checker, and to the rest of the editorial and production staff at PPI—especially Tom Bergstrom, production specialist; Richard Iriye, typesetter; Sam Webster, publishing systems manager; Cathy Schrott, editorial operations manager; Grace Wong, director of editorial operations—thank you all. I very much appreciate your hard work.

To all of you who read this book, it is my greatest hope that you find it useful. Good luck.

Daniel C. Deckler, PhD, PE

Introduction

EXAM FORMAT

The Principles and Practice of Engineering examination (PE exam) for mechanical engineering is an eight-hour exam divided into a morning and an afternoon session. Each session is four hours long and contains 40 problems, or 80 problems in all.

All problems in both the morning and afternoon sessions are multiple-choice. Each problem includes a problem statement that contains all the information needed to find the answer to a question. The problem statement is then followed by four answer options, only one of which is correct. Problems are generally self-contained and independent, so an incorrect choice on one problem typically will not carry over to subsequent problems.

There are three versions of the mechanical engineering exam: thermal and fluids systems, HVAC and refrigeration, and machine design and materials. The mechanical thermal and fluids systems exam covers three major topics: principles (32 problems), hydraulic and fluid applications (24 problems), and energy/power system applications (24 problems.) The knowledge areas and the approximate distribution of problems in the mechanical thermal and fluids systems exam are as follows.

Principles: about 40% of exam problems

- Basic engineering practice: terms, symbols, technical drawings, economics, units
- Fluid mechanics: fluid properties, compressible and incompressible flow
- Principles of heat transfer
- Principles of mass balance
- Thermodynamics: properties, cycles, energy balances, combustion
- Supportive knowledge: pipe system analysis, joints, psychrometrics, codes and standards

Hydraulic and fluid applications: about 30% of exam problems

- Hydraulic and fluid equipment: pumps and fans, compressors, pressure vessels, control valves, actuators, connections
- Distribution systems: pipe flow

Energy/power system applications: about 30% of exam problems

- Energy/power equipment: turbines, boilers, steam generators, internal combustion engines, heat exchangers, cooling towers, condensers
- Cooling/heating: capacity, loads, cycles
- Energy recovery: waste heat, storage
- Combined cycles: components, efficiency

For further information and tips on how to prepare for the mechanical engineering PE exam, consult the *Mechanical Engineering Reference Manual* or the PPI website, **ppi2pass.com**.

THIS BOOK'S ORGANIZATION

Thermal and Fluids Systems Six-Minute Problems is organized into three sections: Principles, Hydraulic and Fluid Applications, and Energy/Power System Applications. Each section contains problems pertaining to the knowledge areas within that division of the NCEES specifications.

Most of the problems in this book are quantitative, meaning that some calculation is necessary to arrive at the correct answer. A few problems are nonquantitative. Some problems will require a little more than six minutes to answer and others a little less. To complete 80 problems in eight hours (480 minutes), you should aim at spending an average of six minutes per problem.

Thermal and Fluids Systems Six-Minute Problems does not include problems related directly to HVAC and refrigeration, nor to machine design and materials. Other books in the *Six-Minute* series provide problems for review in these areas, and the *Mechanical Engineering Reference Manual* provides instruction and example problems in all three areas.

HOW TO USE THIS BOOK

Each problem statement in this book, with its supporting information and answer choices, is presented in the same format as the problems encountered on the PE exam. The solutions explain step by step how the answer is logically derived, to help you follow the reasoning and to provide examples of how you may want to approach your solutions as you take the PE exam.

Each problem includes a hint to provide direction in solving the problem. In addition to the correct solution, you will find an explanation of the faulty reasoning leading to the three incorrect answer choices. The incorrect answers are chosen to show some common mistakes made when solving each type of problem. These may be simple mathematical errors, such as failing to square a term in an equation, or more serious errors, such as using the wrong equation.

To optimize your study time and obtain the maximum benefit from the practice problems, consider the following suggestions.

1. Use the diagnostic exams available at PPI Learning Hub (**ppi2pass.com**) to identify the subjects that you are least familiar with. How you perform on these diagnostic exams will tell you which topics you need to spend more time on and which you can review more lightly.
2. Locate and organize relevant resource materials. (See the References section of this book as a starting point.) As you work problems, some of these resources will emerge as more useful to you than others. These are what you will want to have on hand when taking the PE exam.
3. Work the problems in one subject area at a time, starting with the subject areas that you have the most difficulty with.
4. When possible, work problems without using the hint. Always attempt your own solution before looking at the solutions provided in the book. Use the solutions to check your work or to provide guidance in finding solutions to the more difficult problems. Use the explanations of incorrect answers to help identify pitfalls and to develop strategies to avoid them.
5. Assess your mastery of the subject area with a quiz from Learning Hub's Quiz Generator. If you reach your goal, you're ready to move to the next area. If not, consider some additional study time in this area.
6. Use each subject area's solutions as a guide to understanding general problem-solving approaches. Although problems identical to those presented in this book will not be encountered on the PE exam, the approach to solving problems will be the same.

Solutions presented for each example problem may represent only one of several methods for obtaining a correct answer. Although we have tried to prepare problems with unique solutions, alternative problem-solving methods may occasionally produce a different, but nonetheless appropriate, answer.

References

The minimum recommended library for the mechanical PE exam is PPI's *Mechanical Engineering Reference Manual*. You may also find the following references helpful in completing some of the problems in *Thermal and Fluids Systems Six-Minute Problems*.

American Petroleum Institute. API 570-1998: *Piping Inspection Code: Inspection, Repair, Alteration, and Rerating of In-service Piping Systems.*

______. API 682-2004/ISO 21049: *Pumps—Shaft Sealing Systems for Centrifugal and Rotary Pumps.*

American Society of Mechanical Engineers (ASME). B31.3-2006: *Process Piping.*

______. *Boiler and Pressure Vessel Code.*

Keenan, Joseph H., et al., Steam Tables: *Thermodynamic Properties of Water Including Vapor, Liquid, and Solid Phases.* Krieger Publishing Company.

Shigley, Joseph E., et al., *Mechanical Engineering Design.* McGraw-Hill.

Volk, Michael W., *Pump Characteristics and Applications.* Marcel Dekker.

Nomenclature

a	acceleration	ft/sec^2	m/s^2
a	speed of sound	ft/sec	m/s
A	area	ft^2	m^2
ACFM	actual cubic feet per minute	ft^3/min	n.a.
BHP	boiler or brake horsepower	hp	hp
c	clearance	ft	m
c	distance from neutral axis to extreme fiber	ft	m
c_p	specific heat	Btu/lbm-°F	J/kg·°C
C	attachment factor	–	–
C	cost	\$	\$
C	thermal capacitance	Btu/°F	J/°C
CR	circulation ratio	–	–
CR	corrosion rate	mil/yr	n.a.
d	diameter	ft	m
d	moment arm	ft	m
D	depreciation	\$	\$
E	efficiency	–	–
E	quality factor	–	–
E	specific energy	ft-lbf/lbm	J/kg
f	frequency	Hz	Hz
f	friction factor	–	–
F	force, load	lbf	N
g	gravitational acceleration	ft/sec^2	m/s^2
g_c	gravitational constant	ft-lbm/lbf-sec^2	n.a.
G	mass flow rate per unit area	lbm/ft^2-sec	kg/m^2·s
$G(s)$	transfer function	–	–
h	convection or film coefficient	Btu/hr-ft^2-°F	W/m^2·°C
h	enthalpy	Btu/lbm	J/kg
h	head, head loss, or height	ft	m
h	weld depth	ft	m
h_m	mass transfer coefficient	ft/sec	m/s
H	differential pressure	lbf/ft^2	Pa
H	heat dissipation	Btu-hr/bhp	n.a.
i	effective rate of return	%	%
I	current	A	A
I	mass moment of inertia	ft-lbf-sec^2	m·N·s^2
J	polar area moment of inertia	ft^4	m^4
k	ratio of specific heats	–	–
k	thermal conductivity	Btu/hr-ft-°F	W/m·°C
K	constant	–	–
K	minor loss coefficient	–	–
K_d	density factor	–	–
L	flow resistance	liquid ohm	liquid ohm
L	length	ft	m
L	life	yr	yr
m	mass	lbm	kg
$\dot{m}$	mass flow rate	lbm/hr	kg/h
M	Mach number	–	–
n	number of periods	–	–
n	rotational speed	rpm	rpm
n_s	specific speed	–	–
N	number of poles	–	–
NI	net income	\$	\$
NPSHA	net positive suction head available	ft	m
p	pressure	lbf/ft^2	Pa
P	profit	\$	\$
P	present worth	\$	\$
P	power	hp	W
P	wetted perimeter	ft	m
q	heat	Btu	J
Q	heat flow rate, heat transfer rate	Btu/hr	W
Q	flow rate	gal/sec	L/s
r	radius	ft	m
r	rate	%	%
R	reaction	lbf	kg
R	specific gas constant	ft-lbf/lbm-°R	J/kg·K
RCA	remaining corrosion allowance	ft	m
Re	Reynolds number	–	–
RF	rating factor	–	–

$R(s)$	response function	–	–
s	entropy	Btu/lbm-°F	J/kg·°C
S	Sommerfeld number	–	–
S	stress value	lbf/ft^2	Pa
S_n	expected salvage value in year n	\$	\$
SCFM	standard cubic feet per minute	ft^3/min	n.a.
SG	specific gravity	–	–
Sl	Strouhal number	–	–
t	thickness	ft	m
t	time	sec	s
T	taxes	\$	\$
T	temperature	°F	°C
T	tension	lbf	N
T	torque	ft-lbf	N·m
TU	tower units	–	–
U	overall heat transfer coefficient	Btu/hr-ft-°F	$W/m^2 \cdot °C$
v	velocity	ft/sec	m/s
V	shear force	lbf	N
V	voltage	V	V
V	volume	ft^3	m^3
w	weight per unit length	lbf/ft	N/m
w	width	ft	m
W	mass flow rate	lbm/sec	kg/s
W	weight	lbf	N
W	work	Btu/lbm	J/kg
WHP	water horsepower	hp	hp
x	quality	–	–
z	elevation	ft	m

SYMBOLS

α	diffusivity	ft^2/sec	m^2/s
β	angle of wrap	deg	deg
γ	specific weight	lbf/ft^3	n.a.
ϵ	fractional error	–	–
ϵ	specific roughness	ft	m
η	efficiency	–	–
μ	absolute viscosity	$lbf\text{-}sec/ft^2$	Pa·s
μ	coefficient of friction	–	–
ν	kinematic viscosity	ft^2/sec	m^2/s
ρ	density	lbm/ft^3	kg/m^3
σ	cavitation number	–	–
σ	stress	lbf/ft^2	Pa
τ	shear stress	lbf/ft^2	Pa
υ	specific volume	ft^3/lbm	m^3/kg
ϕ	relative humidity	–	–
ω	angular velocity	rpm	rpm
ω	humidity ratio	–	–

SUBSCRIPTS

ϕ	original, stagnation
A	added (by pump)
add	additional
atm	atmospheric
ave	average
b	base
B	bottom
bend	bending
c	compressible
comb	combustion
corr	corrected
cr	critical
d	design, density
D	discharge
E	extracted (by turbine)
eq	equivalent
f	fin, final, fluid, friction
fg	fluid to gas (vaporization)
g	gas
h	hydraulic
H	high
HHV	higher heating value
i	incompressible, inflow, initial, inner
in	in, input
inj	injection
L	left, low
m	minor, motor
max	maximum
n	nozzle
o	outer
off	off-design
os	oversized
out	out, output
p	pipe, pressure
P	polar
r	radial, reinforcement
R	rated, right
req	required
s	shell, specific, surface
S	suction
sat	saturation
std	standard
t	tangential, turbine
T	tank
th	theoretical, thermal
v	valve
vp	vapor pressure
wv	water vapor
x	horizontal position
y	yield, vertical position
z	potential, position along z-axis
∞	at infinity

1 Principles

PROBLEM 1

An F-14 Tomcat is flying at an altitude of 50,000 ft at 1200 mph. The temperature at this altitude is 393°R. If air is treated as an ideal gas, the Tomcat's Mach number is most nearly

(A) 0.55

(B) 1.6

(C) 1.8

(D) 2.1

Hint: Understand the definition of Mach number.

PROBLEM 2

Airflow velocity over a 1 ft by 1 ft horizontal flat plate is numerically determined to be

$$\mathrm{v}(x, y) = (5 - x)\left(\frac{y}{5} + \frac{y^2}{10} + \frac{y^3}{100}\right)$$

x and y are in feet, and the function $\mathrm{v}(x, y)$ is in feet per second. x is measured from the midpoint of the leading edge of the plate, and y is measured up from the plate. Air pressure is 1 atm and temperature is 70°F.

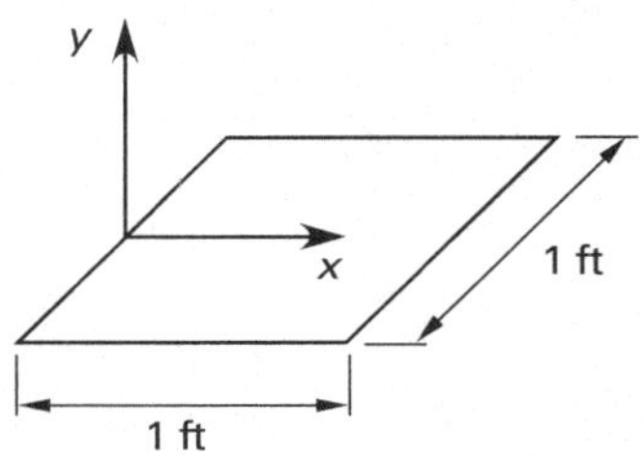

The total shear force acting on the upper surface of the plate is most nearly

(A) 3.4×10^{-7} lbf

(B) 3.8×10^{-7} lbf

(C) 1.1×10^{-5} lbf

(D) 1.2×10^{-5} lbf

Hint: Use Newton's law of viscosity.

PROBLEM 3

Water flows in a horizontal, square cast-iron conduit, 2 in on each side. The flow rate is 0.05 $\mathrm{ft^3/sec}$. The kinematic viscosity of water is 1.08×10^{-5} $\mathrm{ft^2/sec}$, and the density of water is 62.4 $\mathrm{lbm/ft^3}$. Assume all minor losses are incorporated in a minor loss coefficient, K, equal to 0.8. The expected pressure drop over 40 ft of conduit length is most nearly

(A) 0.14 psi

(B) 0.17 psi

(C) 0.19 psi

(D) 0.30 psi

Hint: Use the extended Bernoulli equation.

PROBLEM 4

A 24 in diameter pipeline carries sweet crude oil (specific gravity 0.86, kinematic viscosity 4×10^{-5} $\mathrm{ft^2/sec}$) to a refinery at a rate of 14,000 gal/min. The flow inside the pipe is most nearly

(A) laminar, $\mathrm{Re} = 2.0 \times 10^3$

(B) turbulent, $\mathrm{Re} = 8.5 \times 10^3$

(C) turbulent, $\mathrm{Re} = 41 \times 10^3$

(D) turbulent, $\mathrm{Re} = 500 \times 10^3$

Hint: Do not waste time worrying whether the 24 in diameter is nominal or actual.

PROBLEM 5

Using international standard atmosphere conditions, the maximum pressure on the nose of a plane flying at Mach 1 at sea level can be calculated using the compressible flow theory. Using incompressible flow theory will introduce an error that is most nearly

(A) 10%

(B) 21%

(C) 27%

(D) 220%

Hint: The Bernoulli equation and isentropic tables are useful.

PROBLEM 6

An 800 gpm centrifugal pump generating 700 ft of head is pumping a light, hydrocarbon oil through a pipeline. If the pump is 50% efficient, the maximum possible temperature increase of the oil as it passes through the pump is most nearly

(A) 0.36°F

(B) 0.90°F

(C) 1.8°F

(D) 3.6°F

Hint: The specific heat, c_p, of the oil will be needed.

PROBLEM 7

A boiler generates 200,000 lbm/hr of 85% quality saturated steam at 1500 psia using 220°F feedwater. The energy absorbed by the steam is most nearly

(A) 180 MM Btu/hr

(B) 200 MM Btu/hr

(C) 220 MM Btu/hr

(D) 280 MM Btu/hr

Hint: Think about enthalpies.

PROBLEM 8

In the piping system shown, 100 gal/min of water is pumped from one tank to another.

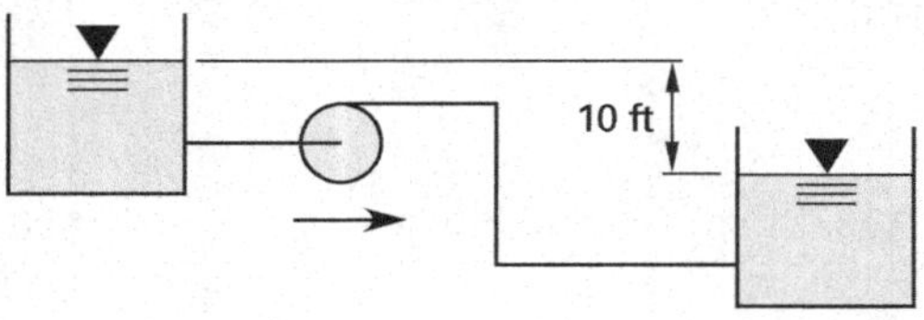

At this flow, the head loss due to piping and fittings is 90 ft. The pump has an efficiency of 60% and is driven by an electric motor with an efficiency of 90%. The fluid density is 50 lbm/ft^3. The power input to the motor is most nearly

(A) 0.38 hp

(B) 3.0 hp

(C) 3.4 hp

(D) 3.8 hp

Hint: Pump power depends on both pump head and flow rate.

PROBLEM 9

A pump can supply 20 ft of head to a water flow of 5 lbm/sec. The water enters from a stagnant lake and exits as a free jet 15 ft above the lake, both locations at atmospheric pressure. The maximum velocity that the 5 lbm/sec flow can have for this single pump is most nearly

(A) 13 ft/sec

(B) 18 ft/sec

(C) 36 ft/sec

(D) 320 ft/sec

Hint: Use the extended Bernoulli equation, and find the *maximum* velocity

PROBLEM 10

The air in one large room is at 70°F, 1.0 atm of pressure, and 80% relative humidity. A 50 ft long duct with a cross-sectional area of 2 ft^2 connects this room to another at 60°F, 1.0 atm, and 10% relative humidity. The diffusivity of water vapor in air is 1.0 ft^2/hr. What is the rate of vapor transfer by diffusion between the two rooms?

(A) 1.3×10^{-5} lbm/hr

(B) 1.7×10^{-5} lbm/hr

(C) 3.4×10^{-5} lbm/hr

(D) 4.3×10^{-5} lbm/hr

Hint: The pressure is uniform.

PROBLEM 11

An intermediate-depth underwater mine designed for coastal defense consists of a hollow, right-circular cylinder that encases an MK42 torpedo. Because of the tether design, the system can be modeled as a double pendulum as shown, with one pivot at the bottom of the case and one at the top of the anchor. The first two natural frequencies of oscillation for this particular system are 0.310 rad/sec and 1.81 rad/sec.

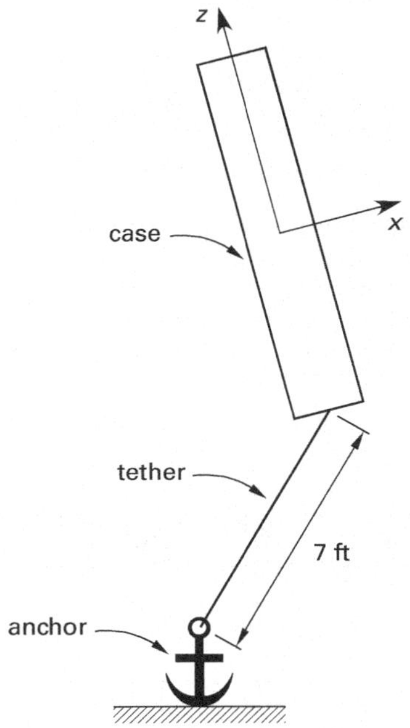

case properties	
length	122 in
outside diameter	23 in
inside diameter	22.25 in
weight	1526 lbf
center of gravity (measured from bottom of case)	71.2 in
I_x	728 ft-lbf-sec^2
I_y	728 ft-lbf-sec^2
I_z	not available

For seawater at 50°F the current velocities (in knots) that will cause the mine to oscillate at its two natural frequencies are most nearly

(A) 0.280 kt, 1.64 kt

(B) 0.473 kt, 2.76 kt

(C) 1.76 kt, 10.3 kt

(D) 3.36 kt, 19.6 kt

Hint: The Reynolds number will be needed, so it will be necessary to assume a reasonable number for the current velocity.

PROBLEM 12

Water is being drawn through a siphon from a tank as shown. The temperature of the tank is 70°F, and the atmospheric pressure is 14.7 psia. The velocity of the water at the lower end of the siphon is 30 ft/sec.

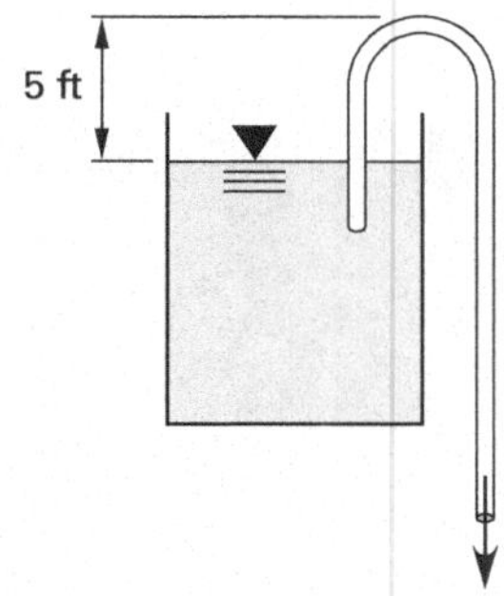

The pressure in the siphon at its highest point is most nearly

(A) 0.44 psia

(B) 6.5 psia

(C) 8.7 psia

(D) 13 psia

Hint: Choose a zero height reference that simplifies the Bernoulli equation.

PROBLEM 13

A 10 in schedule-80 pipe transports saturated water at an initial temperature of 400°F and a flow rate of 7230 lbm/hr. The inside film coefficient can be assumed to be infinite. The outside film coefficient can also be assumed to be infinite, and the outside temperature is 66°F. The pipe is insulated with calcium silicate so that the temperature loss is limited to 1°F per 40 ft of pipe. The minimum thickness of the insulation is most nearly

(A) 3.4 in

(B) 3.8 in

(C) 6.9 in

(D) 8.8 in

Hint: Compared with that of the insulation, the thermal resistance of the steel is negligible.

PROBLEM 14

The air-transfer pipe on an air compressor has an outside diameter of 0.5 in and is made of steel with a thermal conductivity of 27 Btu-ft/hr-ft^2-°F. Under typical operating conditions, the outer surface of the pipe is at a temperature of 150°F and is exposed to ambient air at

80°F, with a convection coefficient of 3 Btu/hr-ft^2-°F. Attached to the pipe are 100 equally spaced annular fins (also known as circular or transverse fins) 0.125 in thick and 1.25 in long.

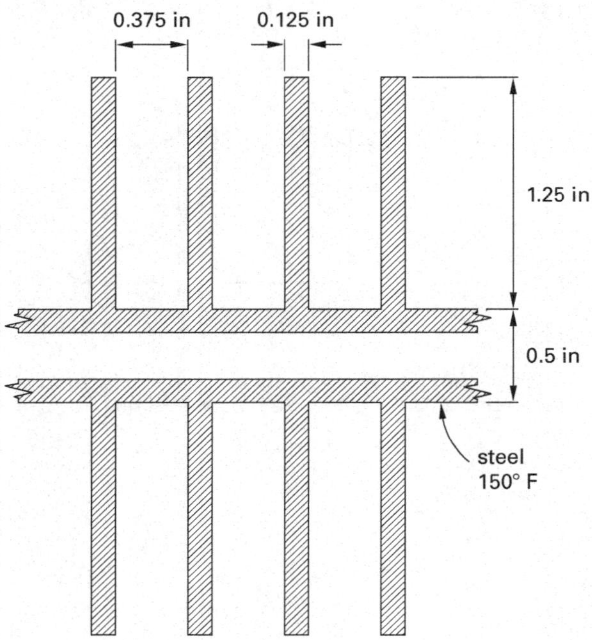

The heat transfer rate from all the fins is most nearly

(A) 870 Btu/hr

(B) 1300 Btu/hr

(C) 1700 Btu/hr

(D) 2200 Btu/hr

Hint: A fin efficiency graph will be helpful.

PROBLEM 15

In the oil supply system shown, oil is pumped from the reservoir to a bearing via the manifold. The oil exiting the bearing is then collected and funneled back into the reservoir.

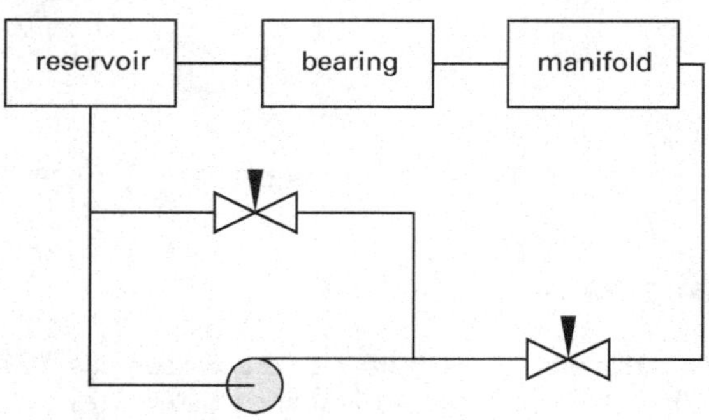

It is desired that the bearing be supplied with oil at 15 psi gauge. The primary functions of the two needle valves are to

I. control the pressure at the pump intake and the pressure at the manifold

II. control the flow rate to the manifold and the pressure at the manifold

III. control the flow rate to the pump intake and the flow rate to the manifold

IV. control the pressure at the pump intake and the flow rate to the manifold

(A) I only

(B) II only

(C) III only

(D) III and IV

Hint: This is a nonquantitative problem. Think about what happens if one needle valve is completely closed while the other is completely open.

PROBLEM 16

A jet of water from a stationary nozzle impinges on a moving block and vane of mass 50 lbm with a turning angle of 30°. The vane is initially at rest against the ground with a coefficient of static friction of 0.1. The water exits the nozzle and hits the vane at 150 ft/sec through an area of 0.005 ft^2.

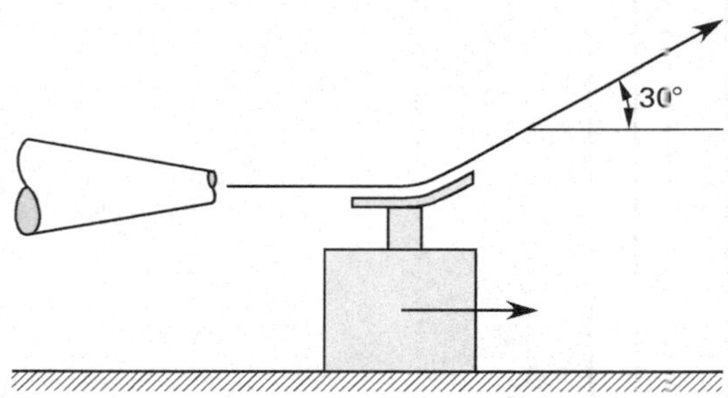

The initial acceleration of the mass is most nearly

(A) 0.27 ft/sec^2

(B) 8.6 ft/sec^2

(C) 19 ft/sec^2

(D) 29 ft/sec^2

Hint: Neglect friction between the vane and fluid.

PROBLEM 17

Air enters a converging-diverging nozzle with a stagnation temperature of 240°F and a stagnation pressure of

200 psia. In the diverging portion of the nozzle, a normal shock is located where M is equal to 1.8. If the exit area is three times larger than the throat area, what is the exit Mach number?

(A) 0.20

(B) 0.25

(C) 2.4

(D) 2.6

Hint: Isentropic flow and normal shock tables are useful.

PROBLEM 18

6 in schedule-40 pipe is used to transport 50,000 lbm/hr of 550°F steam at 500 psia. The pressure drop per 100 ft of pipe is most nearly

(A) 2.2 lbf/in^2

(B) 3.6 lbf/in^2

(C) 6.1 lbf/in^2

(D) 26 lbf/in^2

Hint: A typical friction factor for turbulent flow in steel pipe is 0.02.

PROBLEM 19

A company is considering the purchase of a piece of equipment costing $130,000. If purchased, the equipment will generate $45,000 of net profit for each of the next four years. This net profit represents all profits less operating and maintenance costs, but before taxes. At the end of these four years, the equipment will be salvaged for $10,000. Straight-line depreciation is used. The company's tax rate is 45% and the minimum attractive rate of return is 8%. What is the resulting present value, and should this purchase be made?

(A) −$5500; do not make purchase

(B) −$4000; do not make purchase

(C) $4000; make purchase

(D) $5500; make purchase

Hint: A high present value indicates a more attractive option.

PROBLEM 20

Air flows across a 4 ft^2 area covered with water. The air is at 70°F and 50% relative humidity, while the water is maintained at 60°F. The air velocity is such that the average mass transfer coefficient is 0.10 ft/sec. The rate of evaporation from the water surface is most nearly

(A) 0.091 lbm/hr

(B) 0.37 lbm/hr

(C) 0.47 lbm/hr

(D) 1.1 lbm/hr

Hint: Mass transfer by diffusion is proportional to the difference in vapor density.

PROBLEM 21

Which of the following can be done to increase the net positive suction head available (NPSHA) to a pump?

I. Use a larger diameter pipe.

II. Place a throttling valve in the intake line.

III. Pressurize the supply tank.

IV. Increase the height of the supply tank.

(A) I and IV only

(B) II and IV only

(C) I, II, and IV

(D) I, III, and IV

Hint: This is a nonquantitative question. Start with the index of a fluids book.

PROBLEM 22

Which two of the following are valid pipe fitting classifications?

I. NPT

II. SAE

III. UNC

IV. ASME

(A) I and II

(B) I and III

(C) II and III

(D) III and IV

Hint: Do not confuse screw threads with pipe threads.

PROBLEM 23

The evaporator section of a heat recovery steam generator is made up of a drum boiler that generates the steam, a tube bank heat exchanger, and pipes that carry the water and steam. Preheated water arrives in the drum at 600°F. The drum operates at 1900 psia and produces 95% quality steam at a rate of 540,000 lbm/hr. Each tube in the tube bank has a useful heat-exchange surface of 5.3 ft^2 and an overall coefficient of heat transfer of 218.2 Btu/hr-ft^2-°F. The average temperature difference between the water in the tube bank and flue gas is 32°F. Because of heat losses, only 92% of the heat transmitted to the drum contributes to the generation of steam. The number of tubes needed to maintain the required steam production is most nearly

(A) 6700 tubes

(B) 8000 tubes

(C) 8400 tubes

(D) 260,000 tubes

Hint: Find the heat transfer rate needed to produce the steam.

PROBLEM 24

An axial flow pump running at sea level is used to transfer water at 80°F from an open-air tank to a higher level. The friction loss in the suction pipe is 5 ft. The net positive suction head available is 10 ft. The suction lift is most nearly

(A) 10 ft

(B) 18 ft

(C) 26 ft

(D) 30 ft

Hint: Be sure to use units of "feet absolute."

PROBLEM 25

The smallest nominal schedule-40 pipe diameter that should be used to transport 1750 gpm of water through a city's water supply system is most nearly

(A) 4.0 in

(B) 12 in

(C) 32 in

(D) 48 in

Hint: A rule of thumb will provide a quick estimate.

PROBLEM 26

A boiler horsepower is

I. a unit of power equal to approximately 33,475 Btu/hr, or 9.81 kW

II. equal to one mechanical horsepower

III. the boiler capacity needed to produce one unit of horsepower using a turbine operating at 98% efficiency

IV. the amount of power needed to convert 34.5 lbm/hr of feedwater at 212°F and atmospheric pressure to dry, saturated steam at the same temperature and pressure

(A) I only

(B) I and III

(C) I and IV

(D) II and III

Hint: Boiler horsepower is traditionally used to rate and purchase packaged fire tube boilers.

PROBLEM 27

A boiler produces 200,000 lbm/hr of saturated steam at 1400 psig set pressure. The nozzle area of the steam relief valve is most nearly

(A) 2.7 in^2

(B) 3.1 in^2

(C) 8.9 in^2

(D) 24 in^2

Hint: This problem describes a safety valve.

PROBLEM 28

A nozzle of SA-106 grade B material with a 4.5 in inner diameter abuts a seamless section of a vessel. The vessel has a maximum allowable working pressure of 250 psi and a design temperature of 500°F. The vessel's inside diameter is 68 in and its walls are 0.625 in thick. If the nozzle provides 0.518 in^2 of reinforcement, the amount of additional reinforcement needed is most nearly

(A) 1.0 in^2

(B) 1.2 in^2

(C) 2.7 in^2

(D) 3.4 in^2

Hint: Refer to ASME Sec. VIII.

PROBLEM 29

A section of 3 in, seamless, schedule-40, SA-106 grade B pipe in a crude distillation unit operates at 250 psi and 600°F. The pipe was installed in 1990, and subsequent ultrasonic thickness readings are given in the following table.

year	thickness (in)
1995	0.208
1998	0.200
current year	0.175

The remaining corrosion allowance is most nearly

(A) 0.12 in

(B) 0.15 in

(C) 0.16 in

(D) 0.19 in

Hint: Use ASME code B31.3 and/or API 570.

PROBLEM 30

A full journal bearing with a length/diameter (L/d) ratio of 1, a diameter of 1.5 in, and a radial clearance of 2 mils supports a journal carrying a 495 lbf radial load running at 3600 rpm. If the inlet temperature of the SAE 30 supply oil is 120°F, the temperature rise of the oil is most nearly

(A) 18°F

(B) 26°F

(C) 36°F

(D) 56°F

Hint: Design charts from Raimondi and Boyd will be useful.

PROBLEM 31

A closed feedwater heater subsystem is shown. Refrigerant-12, condensed to a saturated liquid at 0°F and with a quality of zero, is pumped isentropically into the heater at a rate of 1 lbm/sec and a pressure of 100 psia.

Once heated, the R-12 leaves the heater, still as a saturated liquid with a quality of zero and still at a pressure of 100 psia. The energy to heat it comes from R-12 returning in the form of a superheated gas, flowing at a rate of 0.5 lbm/sec, a temperature of 100°F, and a pressure of 120 psia.

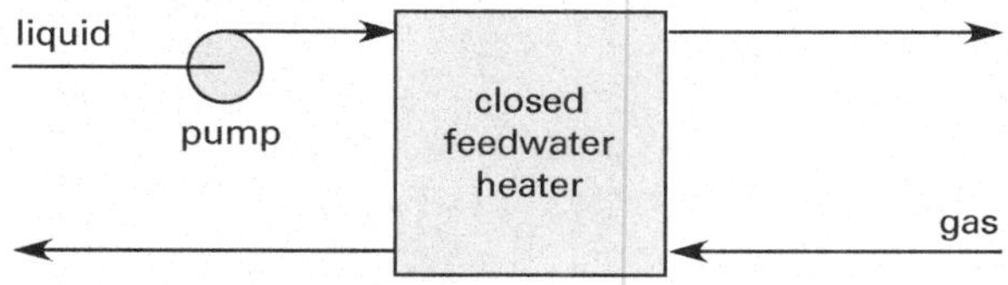

Assume that all the energy lost by the flow of gas is gained by the flow of liquid, that there are no pressure drops in the feedwater heater, and that potential and kinetic energies can be neglected. The quality of the superheated gas as it leaves the feedwater heater is most nearly

(A) 27%

(B) 40%

(C) 84%

(D) 86%

Hint: Use the energy balance to calculate the enthalpy of the gas leaving the heater.

PROBLEM 32

The cooling water requirement for the lube oil system of an 1100 hp gas engine with a water-cooled exhaust manifold is most nearly

(A) 17.6 gpm

(B) 132 gpm

(C) 484 gpm

(D) 1320 gpm

Hint: Use an approximation here.

PROBLEM 33

The quality of 1800 psia saturated steam required to absorb 800 Btu/lbm from a boiler when the feedwater temperature is 280°F is most nearly

(A) 20%

(B) 30%

(C) 80%

(D) 90%

Hint: Do not forget about the energy in the feedwater.

PROBLEM 34

Air flowing at a rate of 1 lbm/sec at 70°F and 14.7 psia is to be compressed to 200 psia. Which of the following statements are true?

I. An isentropic process would require the least amount of power.

II. An adiabatic process must also be an isentropic process.

III. The amount of power required, within 10%, for an isentropic process would be 200 hp.

(A) I only

(B) II only

(C) III only

(D) I and III only

Hint: A typical isothermal process uses intercooling between staged compressors.

PROBLEM 35

A flat, 3 ft diameter, unstayed head is to be designed for a life of 30 years and a design pressure of 150 psi. The maximum allowable stress is 17,500 psi, and the attachment factor is 0.20. If the corrosion rate is 8 mil/yr and the joint efficiency is 1, the design thickness of the head is most nearly

(A) 0.24 in

(B) 0.36 in

(C) 1.7 in

(D) 3.6 in

Hint: Refer to ASME Sec. VIII, Div. I.

PROBLEM 36

An unstayed, circular blind flange is needed to isolate a vessel with an internal design pressure specified at 200 psi. The inside diameter of the vessel is 2 ft and the allowable stress is 13,000 psi. If the attachment factor is 0.17, the required thickness of the blind flange is most nearly

(A) 0.10 in

(B) 1.2 in

(C) 3.0 in

(D) 6.6 in

Hint: Refer to ASME Sec. VIII, Div. I.

PROBLEM 37

A centrifugal feedwater pump has its seals cooled using piping plan 12 to inject water at the seal interface. The heat generation at the seal faces is 1.5 kW. The minimum allowable injection flow rate is most nearly

(A) 0.92 L/min

(B) 1.8 L/min

(C) 3.7 L/min

(D) 7.7 L/min

Hint: Use the American Petroleum Institute standard API 682.

PROBLEM 38

An air-standard diesel engine has a compression ratio of 16:1. The air pressure at the end of the compression process is most nearly

(A) 110 psi

(B) 240 psi

(C) 540 psi

(D) 710 psi

Hint: In an air-standard engine, compression and expansion processes are assumed to be isentropic.

PROBLEM 39

Air is held in a reservoir at 20°C and 500 kPa (absolute). The air exits a hole to the atmosphere with a Mach number most nearly

(A) 0.64

(B) 1.7

(C) 2.1

(D) 4.4

Hint: Treat the hole as a converging-diverging duct.

PROBLEM 40

A fully annealed 1.250 in OD × 0.049 in hydraulic return line made of 304 stainless steel operates between 100°F and 500°F under normal (A) severity of service. 10 gpm flows through the tubing, which is rated for 1000 psi under normal conditions and temperatures. The maximum safe pressure is most nearly

(A) 650 psi

(B) 750 psi

(C) 960 psi

(D) 1000 psi

Hint: Use ASME Standard B31.1.

PROBLEM 41

A fully annealed hydraulic pressure line made of C-1010 steel operates at a maximum pressure of 3000 psi and between 0°F and 200°F. The maximum flow rate is 15 gpm with severity of service normal (A). The lightest acceptable tubing for this application is

(A) 0.375 × 0.049

(B) 0.500 × 0.058

(C) 0.625 × 0.028

(D) 0.750 × 0.083

Hint: Use Parker Hannifin's inch tube pressure rating tables.

PROBLEM 42

The pipe assembly shown is made from schedule-40 pipe. At location C, the nominal diameter of the pipe is 0.75 in, and the static pressure is 40 psi. At locations A and B, the nominal diameter of the pipe is 0.5 in, and the water flows out of the pipe assembly into atmospheric air. At location A, the velocity of the water is 12 ft/sec, and at location B, the velocity of the water is 25 ft/sec. Neglect the weight of the pipe and the water in the pipe. The water is at 60°F.

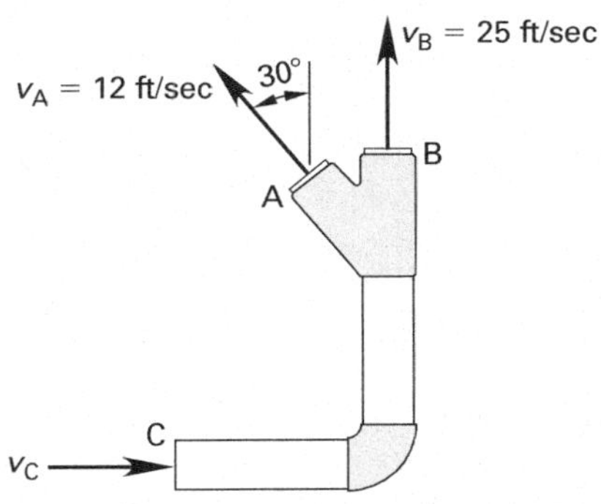

The horizontal force exerted on the elbow at location C needed to hold the pipe assembly in equilibrium is most nearly

(A) 3.5 lbf

(B) 16 lbf

(C) 21 lbf

(D) 25 lbf

Hint: A control volume will be useful here.

PROBLEM 43

The water velocity in a 1 in, schedule-40 pipe with a rapid-closing valve is 3 ft/sec. The water is at 60°F, and the initial pressure in the pipe is 60 psi. Use a safety factor of 1.1. When the valve is closed, the maximum pressure that will result is most nearly

(A) 60 psi

(B) 130 psi

(C) 190 psi

(D) 270 psi

Hint: The speed of sound in water will be needed.

PROBLEM 44

A pump curve is shown.

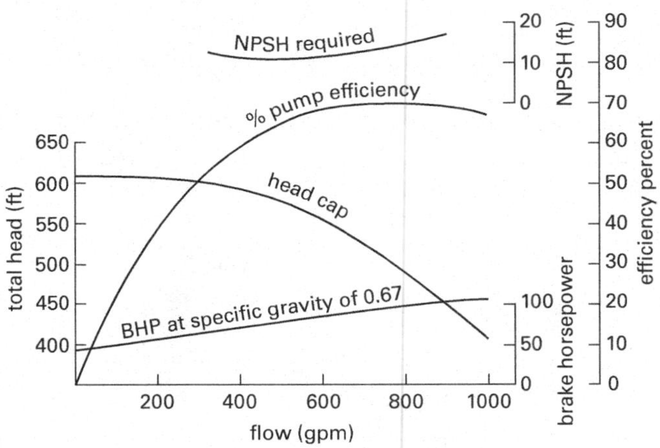

The flow rate at the pump's operating point is 400 gpm. The amount of pump head needed at the pump shaft is most nearly

(A) 840 ft

(B) 980 ft

(C) 1100 ft

(D) 1500 ft

Hint: The operating point is still important.

PROBLEM 45

An ideal Rankine cycle and an incomplete data table are shown.

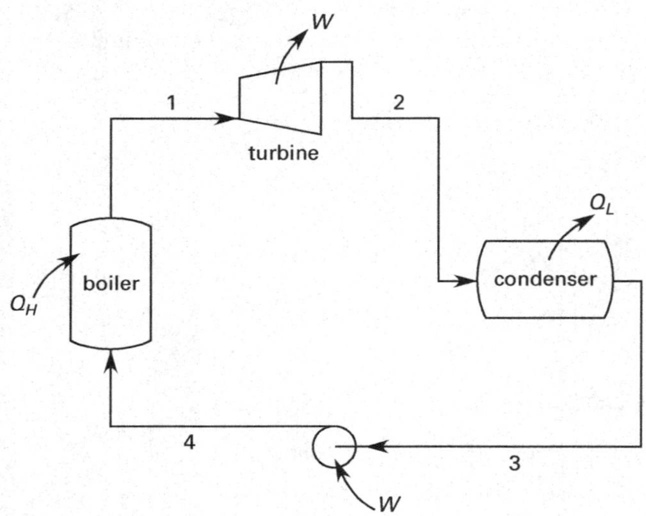

state	pressure (psi)	temperature (°F)	enthalpy (Btu/lbm)	entropy (Btu/lbm-°R)
1	500	700	1357	1.6112
2	2			
3			94	
4			96	

The heat flow per unit mass from the condenser is most nearly

(A) 94 Btu/lbm

(B) 840 Btu/lbm

(C) 1020 Btu/lbm

(D) 1260 Btu/lbm

Hint: The steam quality at state 2 will be needed.

PROBLEM 46

For a combined gas turbine/vapor power cycle, which two statements are true?

I. The thermal efficiency of the combined cycle is greater than the thermal efficiency of either cycle working individually.

II. Gas turbine power cycles typically operate at lower temperatures than steam power cycles.

III. A binary vapor power cycle is classified as a combined power cycle.

IV. The practical upper limit for the efficiency of a combined power cycle is considered to be approximately 60%.

(A) I and III

(B) II and IV

(C) II and III

(D) III and IV

Hint: One of the true statements is not intuitive.

SOLUTION 1

The Mach number of an object is the ratio of the object's speed to the speed of sound in the medium through which it is flying. This definition makes it necessary to determine the speed of sound, a, in air at 50,000 ft. This can be done using the formula

$$a = \sqrt{kg_cRT}$$

Since air is treated as an ideal gas, the ratio of specific heats, k, and the specific gas constant, R, are both constant. For air,

$$\begin{aligned} k &= 1.4 \\ R &= 53.35 \text{ ft-lbf/lbm-°R} \end{aligned}$$

The given temperature at 50,000 ft is

$$T = 393\text{°R}$$

The speed of sound at 50,000 ft is

$$\begin{aligned} a &= \sqrt{kg_cRT} \\ &= \sqrt{(1.4)\left(32.2\ \frac{\text{ft-lbm}}{\text{lbf-sec}^2}\right)\left(53.35\ \frac{\text{ft-lbf}}{\text{lbm-°R}}\right)(393\text{°R})} \\ &= 972 \text{ ft/sec} \end{aligned}$$

The Mach number is

$$\begin{aligned} \text{M} = \frac{\text{v}}{a} &= \frac{\left(1200\ \frac{\text{mi}}{\text{hr}}\right)\left(\frac{1\text{ hr}}{3600\text{ sec}}\right)\left(5280\ \frac{\text{ft}}{\text{mi}}\right)}{972\ \frac{\text{ft}}{\text{sec}}} \\ &= 1.81 \quad (1.8) \end{aligned}$$

The answer is (C).

Why Other Options Are Wrong

(A) This incorrect answer results when the Mach number ratio is inverted.

(B) This incorrect answer results when the speed of sound at sea level (1130 ft/sec) is used to compute the Mach number.

(D) This incorrect answer results when the ratio of specific heats is neglected in calculating the speed of sound.

SOLUTION 2

Newton's law of viscosity applied at the surface (where y equals 0) is

$$\begin{aligned} \tau_{y=0} = \mu\frac{d\text{v}}{dy} &= \mu\left(\frac{\partial \text{v}(x,y)}{\partial y}\right)_{y=0} \\ &= \mu(5-x)\left(\frac{1}{5}+\frac{y}{5}+\frac{3y^2}{100}\right)_{y=0} \\ &= \mu(5-x)\left(\frac{1}{5}\ \text{sec}^{-1}\right) \end{aligned}$$

Air at 70°F has a viscosity (μ) of 3.80×10^{-7} lbf-sec/ft^2. v(x, y) has units of ft/sec, so the derivative of v(x, y) with respect to y has units of sec^{-1}, and τ has units of lbf/ft^2. After integrating, the unit area in square feet can be multiplied into the result, giving an answer in lbf.

The shear force is obtained by integrating the shear stress, τ, over the area of the plate, noting that the differential area, dA, is equal to $dxdz$ and noting that x goes from 0 ft to 1 ft and z goes from −0.5 ft to 0.5 ft.

$$\begin{aligned} F &= \int \tau_{y=0}dA \\ &= \int_{z=-0.5\text{ ft}}^{0.5\text{ ft}}\int_{x=0\text{ ft}}^{1\text{ ft}} \frac{\mu}{5}(5-x)\,dxdz \\ &= \int_{-0.5\text{ ft}}^{0.5\text{ ft}} dz \int_{0\text{ ft}}^{1\text{ ft}} \frac{\mu}{5}(5-x)\,dx \\ &= z\Big|_{-0.5\text{ ft}}^{0.5\text{ ft}} \left(\frac{\mu}{5}\left(5x-\frac{x^2}{2}\right)\right)\Big|_{0\text{ ft}}^{1\text{ ft}} \\ &= 0.9\mu\ \text{ft}^2/\text{sec} \\ &= \left(0.9\ \frac{\text{ft}^2}{\text{sec}}\right)\left(3.80\times 10^{-7}\ \frac{\text{lbf-sec}}{\text{ft}^2}\right) \\ &= 3.42\times 10^{-7}\text{ lbf} \quad (3.4\times 10^{-7}\text{ lbf}) \end{aligned}$$

The answer is (A).

Why Other Options Are Wrong

(B) This incorrect answer results when the shear stress is evaluated for $x = 0$ and $y = 0$ and the result is multiplied by the area of the plate (1 ft^2).

(C) This answer results when an incorrect viscosity of 1.22×10^{-5} lbm/ft-sec is used.

(D) This answer results when the shear stress is evaluated for $x = 0$ and $y = 0$ and an incorrect viscosity of 1.22×10^{-5} lbm/ft-sec is used.

SOLUTION 3

Let point 1 be 40 ft upstream of point 2. The extended Bernoulli equation is

$$\frac{p_1}{\rho} + \frac{v_1^2}{2g_c} + \frac{z_1 g}{g_c} + E_A = \frac{p_2}{\rho} + \frac{v_2^2}{2g_c} + \frac{z_2 g}{g_c} + E_E + E_f + E_m$$

For a horizontal pipe,

$$z_1 = z_2$$

The pipe is continuous, so

$$v_1 = v_2$$

There are no pumps or turbines, so

$$E_A = E_E = 0$$

Solving the extended Bernoulli equation for the drop in pressure, then, gives

$$p_1 - p_2 = \rho(E_f + E_m)$$

E_m represents minor losses and can be calculated from the minor loss coefficient, K.

$$E_m = h_m\left(\frac{g}{g_c}\right) = Kh_v\left(\frac{g}{g_c}\right) = K\left(\frac{v^2}{2g}\right)\left(\frac{g}{g_c}\right) = K\left(\frac{v^2}{2g_c}\right)$$

E_f represents frictional losses and can be evaluated from the equation

$$E_f = h_f\left(\frac{g}{g_c}\right) = \left(\frac{fLv^2}{2dg}\right)\left(\frac{g}{g_c}\right) = \frac{fLv^2}{2dg_c}$$

The average velocity in the pipe, v, is

$$v = \frac{Q}{A} = \frac{0.05\ \frac{\text{ft}^3}{\text{sec}}}{(2\ \text{in})^2\left(\frac{1\ \text{ft}}{12\ \text{in}}\right)^2} = 1.08\ \text{ft/sec}$$

Thus,

$$E_m = K\left(\frac{v^2}{2g_c}\right) = (0.8)\left(\frac{\left(1.80\ \frac{\text{ft}}{\text{sec}}\right)^2}{(2)\left(32.2\ \frac{\text{ft-lbm}}{\text{lbf-sec}^2}\right)}\right) = 4.03 \times 10^{-2}\ \text{ft-lbf/lbm}$$

For a noncircular pipe, the hydraulic diameter, d_h, must be used. For a square duct with side length l,

$$d_h = \frac{4A}{P} = \frac{4l^2}{4l} = l = 2\ \text{in}$$

To find the friction factor, f, the Reynolds number of the flow, Re, must first be found, using the hydraulic diameter.

$$\text{Re} = \frac{vd}{\nu} = \frac{vd_h}{\nu} = \frac{\left(1.80\ \frac{\text{ft}}{\text{sec}}\right)(2\ \text{in})\left(\frac{1\ \text{ft}}{12\ \text{in}}\right)}{1.08 \times 10^{-5}\ \frac{\text{ft}^2}{\text{sec}}} = 2.78 \times 10^4$$

For a cast-iron pipe, the specific roughness, ϵ, is 8.0×10^{-4} ft. The relative roughness is then

$$\frac{\epsilon}{d_h} = \frac{8.0 \times 10^{-4}\ \text{ft}}{(2\ \text{in})\left(\frac{1\ \text{ft}}{12\ \text{in}}\right)} = 4.8 \times 10^{-3}$$

For these values, a Moody friction factor chart gives a friction factor of 0.033. The frictional loss term can now be calculated.

$$E_f = \frac{fLv^2}{2dg_c} = \frac{fLv^2}{2d_h g_c} = \frac{(0.033)(40\ \text{ft})\left(1.80\ \frac{\text{ft}}{\text{sec}}\right)^2}{(2)(2\ \text{in})\left(\frac{1\ \text{ft}}{12\ \text{in}}\right)\left(32.2\ \frac{\text{ft-lbm}}{\text{lbf-sec}^2}\right)} = 3.99 \times 10^{-1}\ \text{lbf-ft/lbm}$$

The pressure drop is

$$\begin{aligned} p_1 - p_2 &= \rho(E_f + E_m) \\ &= \left(62.4\ \frac{\text{lbm}}{\text{ft}^3}\right)\left(\begin{array}{r} 3.99\times 10^{-1}\ \dfrac{\text{ft-lbf}}{\text{lbm}} \\ +\ 4.03\times 10^{-2}\ \dfrac{\text{ft-lbf}}{\text{lbm}} \end{array}\right) \\ &\quad \times\left(\frac{1\ \text{ft}}{12\ \text{in}}\right)^2 \\ &= 1.90\times 10^{-1}\ \text{lbf/in}^2\quad (0.19\ \text{psi}) \end{aligned}$$

The answer is (C).

Why Other Options Are Wrong

(A) This incorrect answer results when the relative roughness is calculated without converting the pipe's hydraulic diameter to feet.

(B) This incorrect answer results when minor losses are neglected.

(D) This incorrect answer results when the area of the pipe is calculated as a circular cross section of diameter 2 in, the hydraulic diameter.

SOLUTION 4

The size of the Reynolds number determines whether the flow is laminar or turbulent.

$$\text{Re} = \frac{\text{v}d}{\nu}$$

The average velocity of the flow inside the pipe is found from the formula

$$\begin{aligned} Q &= \text{v}A \\ \text{v} &= \frac{Q}{A} = \frac{\left(14{,}000\ \dfrac{\text{gal}}{\text{min}}\right)\left(\dfrac{1\ \text{ft}^3}{7.48\ \text{gal}}\right)\left(\dfrac{1\ \text{min}}{60\ \text{sec}}\right)}{\left(\dfrac{\pi}{4}\right)(24\ \text{in})^2\left(\dfrac{1\ \text{ft}}{12\ \text{in}}\right)^2} \\ &= 9.9\ \text{ft/sec} \end{aligned}$$

The Reynolds number is

$$\text{Re} = \frac{\left(9.9\ \dfrac{\text{ft}}{\text{sec}}\right)(24\ \text{in})\left(\dfrac{1\ \text{ft}}{12\ \text{in}}\right)}{4\times 10^{-5}\ \dfrac{\text{ft}^2}{\text{sec}}} = 500\times 10^3$$

The answer is (D).

Why Other Options Are Wrong

(A) This incorrect answer results when a flow rate of 1400 gal/min is mistakenly used, πd is used for the area, and the inches-to-feet conversion is neglected when the flow velocity is calculated.

(B) This incorrect answer results when the number of gallons per minute is divided by 3600 instead of 60 when calculating v.

(C) This incorrect answer results when the inches-to-feet conversion is neglected.

SOLUTION 5

At sea level (0 ft altitude) international standard atmosphere conditions are 518.7°R and 14.7 psia.

For compressible flow at Mach 1, the ratio of static to stagnation pressure, p/p_0, is 0.5283.

Thus, for compressible flow the maximum stagnation pressure is

$$p_{2,c} = \frac{p}{0.5283} = \frac{14.7\ \dfrac{\text{lbf}}{\text{in}^2}}{0.5283} = 27.8\ \text{psi}$$

For incompressible flow, Bernoulli's equation is needed.

$$\frac{p_1}{\rho} + \frac{\text{v}_1^2}{2g_c} + \frac{z_1 g}{g_c} = \frac{p_{2,i}}{\rho} + \frac{\text{v}_2^2}{2g_c} + \frac{z_2 g}{g_c}$$

For a streamline connecting a point away from the plane (1) to the stagnation point at the nose (2), z_1 is equal to z_2, and v_2 is equal to 0. Solving for $p_{2,i}$,

$$p_{2,i} = p_1 + \frac{\rho \text{v}_1^2}{2g_c}$$

p_1 is 14.7 psia. For air, the gas constant, R, is 53.35 ft-lbf/lbm°R. Density can be calculated from the equation

$$\begin{aligned} \rho &= \frac{p}{RT} = \frac{\left(14.7\ \dfrac{\text{lbf}}{\text{in}^2}\right)\left(12\ \dfrac{\text{in}}{\text{ft}}\right)^2}{\left(53.35\ \dfrac{\text{ft-lbf}}{\text{lbm-°R}}\right)(518.7\text{°R})} \\ &= 0.076\ \text{lbm/ft}^3 \end{aligned}$$

To find v_1, only the speed of sound is needed, because the plane is flying at Mach 1.

$$\begin{aligned} v_1 = a &= \sqrt{kg_cRT} \\ &= \sqrt{(1.4)\left(32.2\ \frac{\text{ft-lbm}}{\text{lbf-sec}^2}\right)\left(53.35\ \frac{\text{ft-lbf}}{\text{lbm-°R}}\right)(518.7\text{°R})} \\ &= 1116.9\ \text{ft/sec} \end{aligned}$$

Then,

$$\begin{aligned} p_{2,i} &= p_1 + \frac{\rho v_1^2}{2g_c} \\ &= 14.7\ \frac{\text{lbf}}{\text{in}^2} + \left(\frac{\left(0.076\ \frac{\text{lbm}}{\text{ft}^3}\right)\left(1116.9\ \frac{\text{ft}}{\text{sec}}\right)^2}{(2)\left(32.2\ \frac{\text{ft-lbm}}{\text{lbf-sec}^2}\right)}\right) \\ &\quad \times \left(\frac{1\ \text{ft}}{12\ \text{in}}\right)^2 \\ &= 24.9\ \text{psi} \end{aligned}$$

The fractional error is

$$\begin{aligned} \epsilon &= \frac{|p_{2,c} - p_{2,i}|}{p_{2,c}} \\ &= \frac{\left|27.8\ \frac{\text{lbf}}{\text{in}^2} - 24.9\ \frac{\text{lbf}}{\text{in}^2}\right|}{27.8\ \frac{\text{lbf}}{\text{in}^2}} \\ &= 0.104 \quad (10\%) \end{aligned}$$

The answer is (A).

Why Other Options Are Wrong

(B) This incorrect answer occurs when k is omitted in calculating v_1.

(C) This incorrect answer occurs when ρv_1^2 is divided by g_c instead of $2g_c$.

(D) This incorrect answer occurs when p is multiplied by 0.5283 instead of divided by it.

SOLUTION 6

This problem can be solved by equating the brake horsepower of the pump to the heat flow rate. The brake horsepower of the pump is

$$\text{BHP} = \frac{Q\gamma h}{\eta}$$

But $\dot{m} = Q\gamma$, so the brake horsepower equation can be rewritten as

$$\text{BHP} = \frac{\dot{m}h}{\eta}$$

From heat transfer, the heat flow rate is

$$Q = \dot{m}c_p\Delta T$$

The maximum possible temperature increase will occur when all the brake horsepower goes into heating the fluid. Converting Btu to ft-lbf and equating the heat flow equation to the brake horsepower equation gives

$$\dot{m}c_p\Delta T\left(778\ \frac{\text{ft-lbf}}{\text{Btu}}\right) = \frac{\dot{m}h}{\eta}$$

The specific heat for light, hydrocarbon oil is 0.5 Btu/lbm-°F. Because the specific heat is per pound of mass, the right side of the previous equation must be rewritten

$$\dot{m}c_p\Delta T\left(778\ \frac{\text{ft-lbf}}{\text{Btu}}\right) = \left(\frac{\dot{m}h}{\eta}\right)\left(\frac{g}{g_c}\right)$$

Solving for the temperature rise gives

$$\begin{aligned} \Delta T &= \frac{hg}{\left(778\ \frac{\text{ft-lbf}}{\text{Btu}}\right)c_p\eta g_c} \\ &= \frac{(700\ \text{ft})\left(32.2\ \frac{\text{ft}}{\text{sec}^2}\right)}{\left(778\ \frac{\text{ft-lbf}}{\text{Btu}}\right)\left(0.5\ \frac{\text{Btu}}{\text{lbm-°F}}\right)(0.5)\left(32.2\ \frac{\text{ft-lbm}}{\text{lbf-sec}^2}\right)} \\ &= 3.6\text{°F} \end{aligned}$$

The answer is (D).

Why Other Options Are Wrong

(A) This incorrect answer results when a head of 70 ft is used instead of 700 ft.

(B) This incorrect answer results when, in the calculation of the brake horsepower, the water horsepower is multiplied by the efficiency of the pump instead of divided by it.

(C) This incorrect answer results when the efficiency of the pump is neglected.

SOLUTION 7

The energy absorbed by the steam can be found using

$$Q = \dot{m}\Delta h$$

The enthalpy of the feedwater at 220°F is

$$h_{\text{feed}} = 188.22 \text{ Btu/lbm}$$

The enthalpy of the 85% quality, 1500 psia saturated steam is calculated using the equation

$$h_{\text{steam}} = h_f + xh_{fg}$$

From the steam tables,

$$h_f = 611.5 \text{ Btu/lbm}$$
$$h_{fg} = 557.2 \text{ Btu/lbm}$$

The enthalpy of the steam is

$$\begin{aligned} h_{\text{steam}} &= 611.5\ \frac{\text{Btu}}{\text{lbm}} + (0.85)\left(557.2\ \frac{\text{Btu}}{\text{lbm}}\right) \\ &= 1085 \text{ Btu/lbm} \end{aligned}$$

The energy absorbed by the steam is

$$\begin{aligned} Q &= \left(200{,}000\ \frac{\text{lbm}}{\text{hr}}\right)\left(1085\ \frac{\text{Btu}}{\text{lbm}} - 188.22\ \frac{\text{Btu}}{\text{lbm}}\right) \\ &= 179 \times 10^6 \text{ Btu/hr} \quad (180 \text{ MM Btu/hr}) \end{aligned}$$

The answer is (A).

Why Other Options Are Wrong

(B) This incorrect answer results when the enthalpy of 120°F feedwater is used instead of the enthalpy of 220°F feedwater.

(C) This incorrect answer results when the initial feedwater enthalpy is not subtracted from the final steam enthalpy.

(D) This incorrect answer results when h_g is used instead of h_{fg}.

SOLUTION 8

The head the pump must provide is simply the friction loss minus the elevation change, because the liquid level in the discharge tank is lower than that of the supply tank (apply Bernoulli equation). So, the pump head is

$$\begin{aligned} h_A &= h_f - \Delta z = 90 \text{ ft} - 10 \text{ ft} \\ &= 80 \text{ ft} \end{aligned}$$

The flow converted to cubic feet per minute is

$$Q = \left(100\ \frac{\text{gal}}{\text{min}}\right)\left(\frac{1 \text{ ft}^3}{7.48 \text{ gal}}\right) = 13.4 \text{ ft}^3/\text{min}$$

The required power is

$$\begin{aligned} P &= \left(\frac{\gamma h_A Q}{\eta_{\text{pump}}\eta_{\text{motor}}}\right)\left(\frac{g}{g_c}\right) \\ &= \left(\frac{\left(50\ \frac{\text{lbm}}{\text{ft}^3}\right)(80 \text{ ft})\left(13.4\ \frac{\text{ft}^3}{\text{min}}\right)}{(0.60)(0.90)}\right) \\ &\quad \times \left(\frac{32.2\ \frac{\text{ft}}{\text{sec}^2}}{32.2\ \frac{\text{ft-lbm}}{\text{lbf-sec}^2}}\right)\left(\frac{1 \text{ hp-min}}{33{,}000 \text{ ft-lbf}}\right) \\ &= 3.0 \text{ hp} \end{aligned}$$

The answer is (B).

Why Other Options Are Wrong

(A) This incorrect answer results when friction loss is neglected.

(C) This incorrect answer results when the elevation change is neglected.

(D) This incorrect answer results when the elevation change is added to the friction loss instead of being subtracted.

SOLUTION 9

Define location 1 at the stagnant lake surface and location 2 at the free-jet exit 15 ft higher. The extended Bernoulli equation is

$$\frac{p_1}{\rho} + \frac{\text{v}_1^2}{2g_c} + \frac{z_1 g}{g_c} + E_A = \frac{p_2}{\rho} + \frac{\text{v}_2^2}{2g_c} + \frac{z_2 g}{g_c} + E_E + E_f + E_m$$

At the stagnant lake surface the velocity, v_1, is zero. Let the height at the lake surface, z_1, be zero; then the height at the exit, z_2, is 15 ft. The pressure at the lake surface is atmospheric and the same value as at the free jet's exit, so the pressure terms cancel.

Using these values, the equation becomes

$$E_A = \frac{\text{v}_2^2}{2g_c} + \frac{(15 \text{ ft})g}{g_c} + E_E + E_f + E_m$$

Solving for v_2,

$$v_2 = \sqrt{2g_c\left(E_A - \frac{(15 \text{ ft})g}{g_c} - E_E - E_f - E_m\right)}$$

No energy is extracted in this system (by a turbine, for example), so E_E is zero. E_f and E_m represent losses; to maximize the velocity, these are set to zero. E_A represents energy added by the pump and can be calculated from the pump head, h_A.

$$E_A = \frac{gh_A}{g_c} = \frac{\left(32.2 \ \frac{\text{ft}}{\text{sec}^2}\right)(20 \text{ ft})}{32.2 \ \frac{\text{ft-lbm}}{\text{lbf-sec}^2}} = 20 \text{ ft-lbf/lbm}$$

The maximum exit velocity can now be calculated.

$$\begin{aligned} v_{2,\text{max}} &= \sqrt{\begin{aligned}&(2)\left(32.2 \ \frac{\text{ft-lbm}}{\text{lbf-sec}^2}\right) \\ &\times \left(20 \ \frac{\text{ft-lbf}}{\text{lbm}} - \frac{(15 \text{ ft})\left(32.2 \ \frac{\text{ft}}{\text{sec}^2}\right)}{32.2 \ \frac{\text{ft-lbm}}{\text{lbf-sec}^2}}\right)\end{aligned}} \\ &= 17.9 \text{ ft/sec} \quad (18 \text{ ft/sec}) \end{aligned}$$

The answer is (B).

Why Other Options Are Wrong

(A) This incorrect answer results when the velocities in the Bernoulli equation are divided by g_c instead of $2g_c$.

(C) This incorrect answer results when the 15 ft elevation is omitted.

(D) This incorrect answer results when the square root in the equation for v_2 is omitted.

SOLUTION 10

The mass flow rate of vapor is

$$\dot{m} = A\alpha\left(\frac{\rho_0 - \rho_L}{L}\right)$$

ρ_0 is vapor density at $x = 0$ ft, and ρ_L is vapor density at $x = L$. At each location, the vapor density is

$$\rho = \phi\rho_{\text{sat}}$$

From steam tables, the saturated vapor densities at the given temperatures are

$$\rho_{\text{sat},70^\circ} = 1.15 \times 10^{-3} \text{ lbm/ft}^3$$

$$\rho_{\text{sat},60^\circ} = 0.829 \times 10^{-3} \text{ lbm/ft}^3$$

Therefore,

$$\begin{aligned} \rho_0 &= (0.80)\left(1.15 \times 10^{-3} \ \frac{\text{lbm}}{\text{ft}^3}\right) \\ &= 0.922 \times 10^{-3} \text{ lbm/ft}^3 \\ \rho_L &= (0.10)\left(0.829 \times 10^{-3} \ \frac{\text{lbm}}{\text{ft}^3}\right) \\ &= 0.829 \times 10^{-4} \text{ lbm/ft}^3 \end{aligned}$$

Substituting into the equation for mass flow rate,

$$\begin{aligned} \dot{m} &= (2 \text{ ft}^2)\left(1.0 \ \frac{\text{ft}^2}{\text{hr}}\right) \\ &\quad \times \left(\frac{0.922 \times 10^{-3} \ \frac{\text{lbm}}{\text{ft}^3} - 0.829 \times 10^{-4} \ \frac{\text{lbm}}{\text{ft}^3}}{50 \text{ ft}}\right) \\ &= 3.36 \times 10^{-5} \text{ lbm/hr} \quad (3.4 \times 10^{-5} \text{ lbm/hr}) \end{aligned}$$

The answer is (C).

Why Other Options Are Wrong

(A) This incorrect answer results when relative humidity is neglected.

(B) This incorrect answer results when the cross-sectional area of the duct is omitted from the calculation.

(D) This incorrect answer results when the relative humidity at $x = 0$ is neglected.

SOLUTION 11

When a cylinder is placed in a flowing liquid, it will experience a phenomenon known as vortex shedding. In the flow pattern around the cylinder, vortices are formed and shed from alternating sides of the cylinder. This formation and shedding of vortices from alternating sides creates a regular change in the pressures and subsequent forces acting on the sides of the cylinder. It is this change in side forces that causes the mine to sway back and forth in a plane perpendicular to the current velocity.

The frequency of the vortex shedding is given in terms of the Strouhal number, Sl, which is in turn a function of the Reynolds number. The Strouhal number is defined as

$$\mathrm{Sl} = \frac{fd}{\mathrm{v}_0}$$

The Strouhal number is obtained from a graph of Strouhal versus Reynolds numbers for flow past a cylinder. The vortex shedding frequency, f, will be equal to the natural frequencies given in the statement of the problem. With the cylinder's diameter, d, given, the free-stream velocity v_0 is the quantity to be found.

To find the Strouhal number, the Reynolds number must be calculated. The Reynolds number is defined as

$$\mathrm{Re} = \frac{\mathrm{v}_0 d}{\nu}$$

To calculate the Reynolds number, a current velocity is needed. Looking at the solutions, a reasonable choice is 1.0 kt, which is 1.688 ft/sec. Using the kinematic viscosity of seawater at 50°F, the Reynolds number is

$$\mathrm{Re} = \frac{\left(1.688\ \frac{\mathrm{ft}}{\mathrm{sec}}\right)(23\ \mathrm{in})\left(\frac{1\ \mathrm{ft}}{12\ \mathrm{in}}\right)}{1.5\times 10^{-5}\ \frac{\mathrm{ft}^2}{\mathrm{sec}}} \approx 2.2\times 10^{5}$$

The kinematic viscosity of regular water (1.58×10^{-5} ft²/sec) could have been used, as seawater and water have similar properties. From the graph of Strouhal versus Reynolds numbers for flow past a circular cylinder, the Strouhal number is equal to 0.2.

The velocities can now be calculated using the Strouhal number equation.

$$\begin{aligned}\mathrm{v}_1 &= \frac{f_1 d}{\mathrm{Sl}} \\ &= \left(\frac{\left(0.31\ \frac{\mathrm{rad}}{\mathrm{sec}}\right)(23\ \mathrm{in})}{0.2}\right)\left(\frac{1\ \mathrm{cycle}}{2\pi\mathrm{rad}}\right) \\ &\quad\times\left(\frac{1\ \mathrm{Hz}}{1\ \frac{\mathrm{cycle}}{\mathrm{sec}}}\right)\left(\frac{1\ \mathrm{ft}}{12\ \mathrm{in}}\right)\left(\frac{1\ \mathrm{kt}}{1.688\ \frac{\mathrm{ft}}{\mathrm{sec}}}\right) \\ &= 0.280\ \mathrm{kt}\end{aligned}$$

$$\begin{aligned}\mathrm{v}_2 &= \frac{f_2 d}{\mathrm{Sl}} \\ &= \left(\frac{\left(1.81\ \frac{\mathrm{rad}}{\mathrm{sec}}\right)(23\ \mathrm{in})}{0.2}\right)\left(\frac{1\ \mathrm{cycle}}{2\pi\mathrm{rad}}\right) \\ &\quad\times\left(\frac{1\ \mathrm{Hz}}{1\ \frac{\mathrm{cycle}}{\mathrm{sec}}}\right)\left(\frac{1\ \mathrm{ft}}{12\ \mathrm{in}}\right)\left(\frac{1\ \mathrm{kt}}{1.688\ \frac{\mathrm{ft}}{\mathrm{sec}}}\right) \\ &= 1.64\ \mathrm{kt}\end{aligned}$$

The answer is (A).

Why Other Options Are Wrong

(B) This incorrect solution results when feet per second are not converted to knots.

(C) This incorrect solution results from not converting f_1 and f_2 from radians per second to hertz.

(D) This incorrect solution results from not converting the diameter from inches to feet.

SOLUTION 12

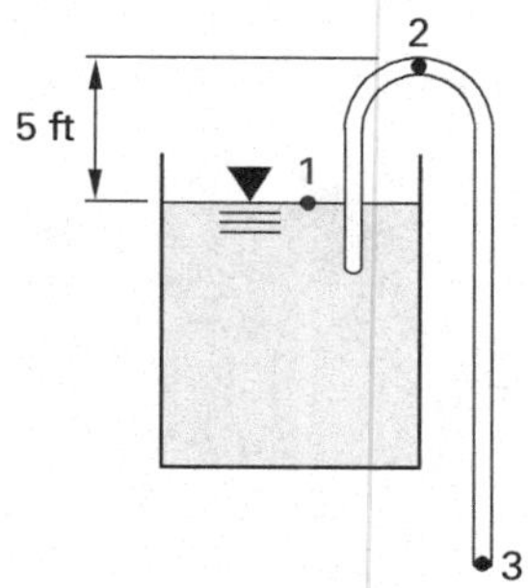

The total energy is constant, so the Bernoulli equation is

$$\frac{p_1}{\rho} + \frac{\mathrm{v}_1^2}{2g_c} + \frac{z_1 g}{g_c} = \frac{p_2}{\rho} + \frac{\mathrm{v}_2^2}{2g_c} + \frac{z_2 g}{g_c}$$

From the law of conservation of mass for an incompressible fluid, the velocity at point 2 is the same as the velocity at point 3. Assuming the velocity at point 1 is 0 ft/sec, and using point 1 as the zero height reference,

$$\frac{p_1}{\rho} + 0 + 0 = \frac{p_2}{\rho} + \frac{\mathrm{v}_2^2}{2g_c} + \frac{z_2 g}{g_c}$$

Solving for the pressure at point 2 yields

$$p_2 = p_1 - \left(\frac{\mathrm{v}_2^2}{2g_c} + \frac{z_2 g}{g_c}\right)\rho$$

The density of water at 70°F is 62.3 lbm/ft³.

$$p_2 = 14.7\ \frac{\text{lbf}}{\text{in}^2} - \left(\frac{\left(30\ \frac{\text{ft}}{\text{sec}}\right)^2}{(2)\left(32.2\ \frac{\text{ft-lbm}}{\text{lbf-sec}^2}\right)} + \frac{(5\ \text{ft})\left(32.2\ \frac{\text{ft}}{\text{sec}^2}\right)}{32.2\ \frac{\text{ft-lbm}}{\text{lbf-sec}^2}} \right) \times \left(62.3\ \frac{\text{lbm}}{\text{ft}^3}\right)\left(\frac{1\ \text{ft}}{12\ \text{in}}\right)^2$$
$$= 6.5\ \text{lbf/in}^2 \quad (6.5\ \text{psia})$$

The answer is (B).

Why Other Options Are Wrong

(A) This incorrect solution results from dividing by g_c instead of $2g_c$ in the kinetic energy term.

(C) This incorrect solution results from omitting the potential energy term zg/g_c.

(D) This incorrect solution results from omitting the kinetic energy term $v^2/2g_c$.

SOLUTION 13

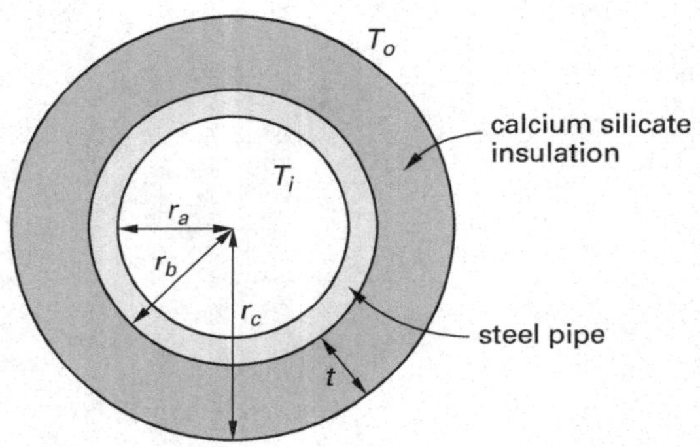

The heat transfer through the pipe is given by the equation

$$Q = \frac{2\pi L(T_i - T_o)}{\frac{1}{r_a h_a} + \frac{\ln \frac{r_b}{r_a}}{k_{\text{pipe}}} + \frac{1}{r_b h_b} + \frac{\ln \frac{r_c}{r_b}}{k_{\text{insul}}} + \frac{1}{r_c h_c}}$$

The outer radius of a 10 in schedule-80 pipe is

$$r_b = \frac{d_o}{2} = \frac{10.75\ \text{in}}{2} = 5.375\ \text{in}$$

The temperature drop in the pipe wall is negligible, and all film coefficients are infinite, so the heat loss through the pipe is

$$Q = \frac{2\pi L(T_i - T_o)k_{\text{insul}}}{\ln \frac{r_c}{r_b}}$$

The outer radius of the insulation is

$$r_c = r_b + t$$

Therefore,

$$\frac{r_c}{r_b} = \frac{r_b + t}{r_b} = 1 + \frac{t}{r_b}$$

Substituting and solving for the thickness gives

$$Q = \frac{2\pi L(T_i - T_o)k_{\text{insul}}}{\ln\left(1 + \frac{t}{r_b}\right)}$$

$$t = r_b\left(e^{\frac{2\pi L(T_i - T_o)}{Q}k_{\text{insul}}} - 1\right)$$

The heat rate required to reduce the water temperature by 1°F is

$$Q = \dot{m}c_p\Delta T = \left(7230\ \frac{\text{lbm}}{\text{hr}}\right)\left(1.08\ \frac{\text{Btu}}{\text{lbm-°F}}\right)(1\text{°F}) = 7810\ \text{Btu/hr}$$

The required thickness is

$$t = (5.375\ \text{in})\left(e^{\frac{(2\pi)(40\ \text{ft})(400\text{°F} - 66\text{°F})\left(0.046\ \frac{\text{Btu-ft}}{\text{hr-ft}^2\text{-°F}}\right)}{7810\ \frac{\text{Btu}}{\text{hr}}}} - 1\right) = 3.44\ \text{in} \quad (3.4\ \text{in})$$

The answer is (A).

Why Other Options Are Wrong

(B) This incorrect answer results when a specific heat equal to 1.00 is used instead of 1.08.

(C) This incorrect answer results when the diameter is used instead of the radius in calculating the thickness.

(D) This incorrect answer results when the subtraction of the 1 in thickness is omitted.

SOLUTION 14

The heat transfer rate from one fin is

$$Q = \eta_f h A_f (T_b - T_\infty)$$

To find the fin area, A_f, the corrected fin length is needed.

$$\begin{aligned} L_{\text{corr}} &= L + \frac{t}{2} \\ &= 1.25 \text{ in} + \frac{0.125 \text{ in}}{2} \\ &= 1.3125 \text{ in} \end{aligned}$$

Now the fin area can be calculated.

$$\begin{aligned} A_f &= 2\pi\left((L_{\text{corr}} + r_1)^2 - r_1^2\right) \\ &= 2\pi\left((1.3125 \text{ in} + 0.25 \text{ in})^2 - (0.25 \text{ in})^2\right)\left(\frac{1 \text{ ft}}{12 \text{ in}}\right)^2 \\ &= 0.104 \text{ ft}^2 \end{aligned}$$

The fin efficiency is found using a fin efficiency graph. The graph parameters are

$$\begin{aligned} \frac{L_{\text{corr}} + r_1}{r_1} &= \frac{1.3125 \text{ in} + 0.25 \text{ in}}{0.25 \text{ in}} = 6.25 \\ L_{\text{corr}}\sqrt{\frac{h}{kt}} &= (1.3125 \text{ in})\left(\frac{1 \text{ ft}}{12 \text{ in}}\right) \\ &\quad \times \sqrt{\frac{3 \ \frac{\text{Btu}}{\text{hr-ft}^2\text{-}^\circ\text{F}}}{\left(27 \ \frac{\text{Btu}}{\text{hr-ft-}^\circ\text{F}}\right)(0.125 \text{ in})\left(\frac{1 \text{ ft}}{12 \text{ in}}\right)}} \\ &= 0.357 \end{aligned}$$

From the graph, the fin efficiency is approximately

$$\eta_f = 0.8$$

The heat transfer rate per fin can now be found.

$$\begin{aligned} Q_{\text{fin}} &= \eta_f h A_f (T_b - T_\infty) \\ &= (0.8)\left(3 \ \frac{\text{Btu}}{\text{hr-ft}^2\text{-}^\circ\text{F}}\right)(0.104 \text{ ft}^2)(150^\circ\text{F} - 80^\circ\text{F}) \\ &= 17.47 \text{ Btu/hr} \end{aligned}$$

The total heat transfer rate of all 100 fins is

$$\begin{aligned} Q_{\text{total}} &= (\text{no. of fins})Q_{\text{fin}} \\ &= (100 \text{ fins})\left(17.47 \ \frac{\text{Btu}}{\text{hr-fin}}\right) \\ &= 1747 \text{ Btu/hr} \quad (1700 \text{ Btu/hr}) \end{aligned}$$

The answer is (C).

Why Other Options Are Wrong

(A) This incorrect answer results from neglecting the inches-to-feet conversion in calculating one of the graph parameters.

(B) This incorrect answer results from using $2\pi L_{\text{corr}}^2$ for the fin area.

(D) This incorrect answer results from neglecting to use the fin efficiency.

SOLUTION 15

The needle valve in the line to the manifold controls the pressure to the manifold. The other needle valve allows oil to leave the line to the manifold and return to the pump intake line; therefore, this valve primarily controls the amount of oil flowing to the manifold.

The answer is (B).

Why Other Options Are Wrong

(A) This answer is incorrect because neither needle valve primarily controls the pressure at the pump intake.

(C) This answer is incorrect because neither needle valve primarily controls the flow rate to the pump intake.

(D) This answer is incorrect because neither needle valve primarily controls the flow rate to the pump intake.

SOLUTION 16

The acceleration in the x-direction can be found with the equation

$$a_x = \frac{F_x g_c}{m_{\text{block}}}$$

Since m_{block} is known, F_x is needed. There are two forces in the x-direction: one due to the water, and the other due to friction.

The force that the block and vane exert on the water is given by the formula

$$F_{x,\text{block}} = \frac{\dot{m}_{\text{water}}(\text{v}_2 \cos\theta - \text{v}_1)}{g_c}$$

The mass flow rate is given by

$$\begin{aligned}\dot{m}_{\text{water}} &= \rho \text{v} A \\ &= \left(62.4\ \frac{\text{lbm}}{\text{ft}^3}\right)\left(150\ \frac{\text{ft}}{\text{sec}}\right)(0.005\ \text{ft}^2) \\ &= 46.8\ \text{lbm/sec}\end{aligned}$$

The velocity into and out of the vane is the same.

$$\text{v}_2 = \text{v}_1 = 150\ \text{ft/sec}$$

Substituting,

$$\begin{aligned}F_{x,\text{block}} &= \frac{\left(46.8\ \frac{\text{lbm}}{\text{sec}}\right) \times \left(\left(150\ \frac{\text{ft}}{\text{sec}}\right)\cos 30^\circ - 150\ \frac{\text{ft}}{\text{sec}}\right)}{32.2\ \frac{\text{ft-lbm}}{\text{lbf-sec}^2}} \\ &= -29.2\ \text{lbf}\end{aligned}$$

The force that the water exerts on the block and vane is equal and opposite to this.

$$\begin{aligned}F_{x,\text{water}} &= -F_{x,\text{block}} \\ &= 29.2\ \text{lbf}\end{aligned}$$

For the frictional force, the y-force due to the water is needed.

$$\begin{aligned}F_{y,\text{water}} &= \frac{\dot{m}_{\text{water}}\text{v}_2 \sin\theta}{g_c} \\ &= \frac{\left(46.8\ \frac{\text{lbm}}{\text{sec}}\right)\left(150\ \frac{\text{ft}}{\text{sec}}\right)\sin 30^\circ}{32.2\ \frac{\text{ft-lbm}}{\text{lbf-sec}^2}} \\ &= 109\ \text{lbf}\end{aligned}$$

There is also the y-force due to the weight of the vane and block.

$$\begin{aligned}F_{y,\text{block}} &= \frac{m_{\text{block}}g}{g_c} \\ &= \frac{(50\ \text{lbm})\left(32.2\ \frac{\text{ft}}{\text{sec}^2}\right)}{32.2\ \frac{\text{ft-lbm}}{\text{lbf-sec}^2}} \\ &= 50\ \text{lbf}\end{aligned}$$

The total force in the vertical direction is

$$F_y = F_{y,\text{water}} + F_{y,\text{block}} = 109\ \text{lbf} + 50\ \text{lbf} = 159\ \text{lbf}$$

The x-force due to friction can be found from the equation

$$F_{x,\text{friction}} = \mu_f F_y = (0.1)(159\ \text{lbf}) = 15.9\ \text{lbf}$$

Friction acts against the water force, so the resulting acceleration is

$$\begin{aligned}a_x &= \frac{F_x g_c}{m} = \frac{(F_{x,\text{water}} - F_{x,\text{friction}})g_c}{m_{\text{block}}} \\ &= \frac{(29.2\ \text{lbf} - 15.9\ \text{lbf})\left(32.2\ \frac{\text{ft-lbm}}{\text{lbf-sec}^2}\right)}{50\ \text{lbm}} \\ &= 8.57\ \text{ft/sec}^2 \quad (8.6\ \text{ft/sec}^2)\end{aligned}$$

The answer is (B).

Why Other Options Are Wrong

(A) This incorrect answer results when the gravitational constant, g_c, is neglected.

(C) This incorrect answer results when friction is neglected.

(D) This incorrect answer results when the frictional force is added to the water force instead of subtracted from it.

SOLUTION 17

The flow is isentropic up to and after the normal shock. From the normal shock table, for M_x equal to 1.8, M_y is equal to 0.62. To determine the properties of the isentropic flow from there to the exit, the ratio of the exit area to the area at which the flow would attain Mach 1 must be known. Note that before the shock, this area (A_x^*) is the throat area. After the shock, this area (A_y^*) is larger. From the shock tables again, for M_x equal to 1.8,

$$\frac{A_x^*}{A_y^*} = \frac{p_{0,y}}{p_{0,x}} = 0.81$$

The needed ratio, A_{exit}/A_y^*, can be calculated as follows.

$$\begin{aligned}\frac{A_{\text{exit}}}{A_y^*} &= \left(\frac{A_{\text{exit}}}{A_x^*}\right)\left(\frac{A_x^*}{A_y^*}\right) = \left(\frac{A_{\text{exit}}}{A_{\text{throat}}}\right)\left(\frac{A_x^*}{A_y^*}\right)\\ &= (3)(0.81)\\ &= 2.4\end{aligned}$$

From the isentropic tables, this ratio corresponds to a Mach number of 0.25.

The answer is (B).

Why Other Options Are Wrong

(A) This answer is incorrect because the change in A^* across the shock is not accounted for.

(C) This incorrect answer is the supersonic solution to the area ratio.

(D) This incorrect answer results when the change in A^* across the shock is not accounted for and the supersonic solution is found.

SOLUTION 18

Method 1

The change in pressure can be found using the equation

$$\Delta p = \rho h_f\left(\frac{g}{g_c}\right) = \left(\frac{h_f}{v}\right)\left(\frac{g}{g_c}\right)$$

The head loss due to friction is

$$h_f = \frac{fL\text{v}^2}{2dg}$$

The friction factor, f, is found using a friction factor chart. To use this chart, however, the Reynolds number is needed, and to compute the Reynolds number it is necessary to have the viscosity of the steam. Since the viscosity of steam is difficult to find, a typical friction factor for turbulent flow in steel pipe of 0.02 will be used.

The steam velocity is

$$\text{v} = \frac{Q}{A} = \frac{\dot{m}}{\rho A} = \frac{\dot{m}v}{A}$$

From the steam tables the specific volume is

$$v = 1.0792\ \text{ft}^3/\text{lbm}$$

6 in schedule-40 pipe has an internal diameter and an internal area of

$$\begin{aligned}d_i &= 6.065\ \text{in}\\ A &= \pi\left(\frac{d_i}{2}\right)^2 = \pi\left(\frac{6.065\ \text{in}}{2}\right)^2 = 28.89\ \text{in}^2\end{aligned}$$

Therefore,

$$\begin{aligned}\text{v} &= \frac{\dot{m}v}{A} = \frac{\left(50{,}000\ \frac{\text{lbm}}{\text{hr}}\right)\left(\frac{1\ \text{hr}}{3600\ \text{sec}}\right)\left(1.0792\ \frac{\text{ft}^3}{\text{lbm}}\right)}{(28.89\ \text{in}^2)\left(\frac{1\ \text{ft}}{12\ \text{in}}\right)^2}\\ &= 74.7\ \text{ft/sec}\end{aligned}$$

The head loss due to friction is

$$\begin{aligned}h_f &= \frac{fL\text{v}^2}{2dg} = \frac{(0.02)(100\ \text{ft})\left(74.7\ \frac{\text{ft}}{\text{sec}}\right)^2}{(2)(6.065\ \text{in})\left(\frac{1\ \text{ft}}{12\ \text{in}}\right)\left(32.2\ \frac{\text{ft}}{\text{sec}^2}\right)}\\ &= 343\ \text{ft}\end{aligned}$$

The pressure drop is

$$\begin{aligned}\Delta p &= \left(\frac{h_f}{v}\right)\left(\frac{g}{g_c}\right)\\ &= \left(\frac{(343\ \text{ft})\left(12\ \frac{\text{in}}{\text{ft}}\right)}{\left(1.0792\ \frac{\text{ft}^3}{\text{lbm}}\right)\left(12\ \frac{\text{in}}{\text{ft}}\right)^3}\right)\left(\frac{32.2\ \frac{\text{ft}}{\text{sec}^2}}{32.2\ \frac{\text{ft-lbm}}{\text{lbf-sec}^2}}\right)\\ &= 2.2\ \text{lbf/in}^2\end{aligned}$$

The answer is (A).

Method 2

Steam viscosity data can be found in the Crane Company's technical paper no. 410. For steam at 550°F and 500 psia, the viscosity in centipoises is

$$\mu = 0.02\ \text{cP}$$

Converting the viscosity to English units gives

$$\begin{aligned}\mu &= (0.02\ \text{cP})\left(\frac{1\ \text{P}}{100\ \text{cP}}\right)\left(\frac{0.002089\ \frac{\text{lbf-sec}}{\text{ft}^2}}{P}\right)\\ &= 4.18\times 10^{-7}\ \text{lbf-sec/ft}^2\end{aligned}$$

Now the Moody friction factor chart can be used to find the friction factor. To use this chart, the Reynolds number and the relative roughness are needed. When the mass flow rate is given, the Reynolds number can be expressed in terms of the mass flow rate per unit area, G.

$$\begin{aligned}G &= \rho \text{v} = \frac{\dot{m}}{A} = \frac{\left(50{,}000\ \frac{\text{lbm}}{\text{hr}}\right)\left(\frac{1\ \text{hr}}{3600\ \text{sec}}\right)}{(28.89\ \text{in}^2)\left(\frac{1\ \text{ft}}{12\ \text{in}}\right)^2}\\ &= 69.2\ \text{lbm/ft}^2\text{-sec}\end{aligned}$$

$$\begin{aligned}\text{Re} &= \frac{dG}{g_c\mu}\\ &= \frac{(6.065\ \text{in})\left(\frac{1\ \text{ft}}{12\ \text{in}}\right)\left(69.2\ \frac{\text{lbm}}{\text{ft}^2\text{-sec}}\right)}{\left(32.2\ \frac{\text{ft-lbm}}{\text{lbf-sec}^2}\right)\left(4.18\times 10^{-7}\ \frac{\text{lbf-sec}}{\text{ft}^2}\right)}\\ &= 2.6\times 10^6\end{aligned}$$

The relative roughness is given by ϵ/d. The specific roughness, ϵ, can be found from tables, and for steel, ϵ equals 0.0002 ft. The relative roughness is

$$\frac{\epsilon}{d} = \frac{0.0002\ \text{ft}}{(6.065\ \text{in})\left(\frac{1\ \text{ft}}{12\ \text{in}}\right)} = 4\times 10^{-4}$$

From the Moody friction factor chart, the friction factor is

$$f = 0.016$$

The head loss due to friction is

$$\begin{aligned}h_f &= \frac{fL\text{v}^2}{2dg} = \frac{(0.016)(100\ \text{ft})\left(74.7\ \frac{\text{ft}}{\text{sec}}\right)^2}{(2)(6.065\ \text{in})\left(\frac{1\ \text{ft}}{12\ \text{in}}\right)\left(32.2\ \frac{\text{ft}}{\text{sec}^2}\right)}\\ &= 274\ \text{ft}\end{aligned}$$

The pressure drop is

$$\begin{aligned}\Delta p &= \left(\frac{h_f}{v}\right)\left(\frac{g}{g_c}\right)\\ &= \left(\frac{(274\ \text{ft})\left(12\ \frac{\text{in}}{\text{ft}}\right)}{\left(1.0792\ \frac{\text{ft}^3}{\text{lbm}}\right)\left(12\ \frac{\text{in}}{\text{ft}}\right)^3}\right)\left(\frac{32.2\ \frac{\text{ft}}{\text{sec}^2}}{32.2\ \frac{\text{ft-lbm}}{\text{lbf-sec}^2}}\right)\\ &= 1.76\ \text{lbf/in}^2\quad (2.2\ \text{lbf/in}^2)\end{aligned}$$

The answer is (A).

Method 3

The Crane Company's technical paper no. 410 also gives the following equation for calculating the pressure drop. This combines all the previous equations and conversion factors into one equation.

$$\Delta p = 0.000336\left(\frac{f\dot{m}^2 v}{d^5}\right)$$

For this equation to be dimensionally correct, the constant must have the following units.

$$\Delta p = \left(0.000336\ \frac{\text{lbf-in}^3\text{-hr}^2}{\text{lbm-ft}^3}\right)\left(\frac{f\dot{m}^2 v}{d^5}\right)$$

To use this equation, the mass flow rate, $\dot{m}$, must be in lbm/hr, the specific volume, v, must be in ft^3/lbm, and the diameter, d, must be in inches.

$$\begin{aligned}\Delta p &= \left(0.000336\ \frac{\text{lbf-in}^3\text{-hr}^2}{\text{lbm-ft}^3}\right)\\ &\quad\times\left(\frac{(0.02)\left(50{,}000\ \frac{\text{lbm}}{\text{hr}}\right)^2\left(1.0792\ \frac{\text{ft}^3}{\text{lbm}}\right)}{(6.065\ \text{in})^5}\right)\\ &= 2.2\ \text{lbf/in}^2\end{aligned}$$

(Using $f = 0.018$ in this equation yields $\Delta p = 2.0$ psi.)

The answer is (A).

Method 4

The Babcock formula for pressure drop when steam flows in a pipe is

$$\begin{aligned}\Delta p_{\text{psi}} &= (0.470)\left(\frac{d_{\text{in}} + 3.6}{d_{\text{in}}^6}\right)(\dot{m}_{\text{lbm/sec}})^2 L_{\text{ft}} v_{\text{ft}^3/\text{lbm}} \\ &= (0.470)\left(\frac{6.065 + 3.6}{(6.065)^6}\right)\left(\frac{50{,}000}{3600}\right)^2 \\ &\quad \times (100)(1.0792) \\ &= 1.9 \\ \Delta p &= 1.9 \text{ psi} \quad (2.2 \text{ lbf/in}^2)\end{aligned}$$

Method 5

The Crane Company's technical paper no. 410 also has a nomograph that can be used to solve this problem graphically. To use their nomograph follow these steps.

1. Locate 50 on the W axis, noting that the scale is in 1000 lbm/hr. Label this point A.
2. Locate the nominal diameter of 6 in on the d axis, and label this point B.
3. Draw a straight line through points A and B. Find the intersection of this line and the Index 2 axis, and label it point C.
4. Locate 0.02 on the friction factor f axis and label it point D.
5. Draw a straight line between points C and D. Find the intersection of this line and the Index 1 axis, and label it point E.
6. Locate 1.0792 ft^3/lbm on the specific volume axis, and label this point F.
7. Draw a straight line between points E and F. The intersection of this line and the ΔP_{100} axis is the change in pressure, 2 psi.

The answer is (A).

Why Other Options Are Wrong

(B) This answer results when the wrong internal pipe diameter is read from the schedule-40 table.

(C) This incorrect answer results when a specific volume of 1.792 ft^3/lbm, instead of 1.0792 ft^3/lbm, is used in the velocity calculation.

(D) This incorrect answer results when the conversion factor for v is squared instead of cubed.

SOLUTION 19

The straight-line depreciation for each year is given by

$$D = \frac{C - S_n}{n} = \frac{\$130{,}000 - \$10{,}000}{4} = \$30{,}000$$

Taxes are owed on net profits less depreciation, so

$$T = r_{\text{tax}}(P - D) = (0.45)(\$45{,}000 - \$30{,}000) = \$6750$$

Then the net income for each year is

$$\text{NI} = P - T = \$45{,}000 - \$6750 = \$38{,}250$$

The present worth of the equipment is the sum of the present worths of the net annual income and the salvage value, less the cost.

$$\begin{aligned}P_{\text{income}} &= (\text{NI})(P/A, i, n) \\ &= (\$38{,}250)(P/A, 8\%, 4) \\ &= (\$38{,}250)(3.3121) \\ &= \$126{,}688 \\ P_{\text{salvage}} &= S(P/F, i, n) \\ &= (\$10{,}000)(P/F, 8\%, 4) \\ &= (\$10{,}000)(0.7350) \\ &= \$7350 \\ P &= P_{\text{income}} + P_{\text{salvage}} - C \\ &= \$126{,}688 + \$7350 - \$130{,}000 \\ &= \$4038 \quad (\$4000)\end{aligned}$$

The present value is greater than zero, so the purchase should be made.

The answer is (C).

Why Other Options Are Wrong

(A) This incorrect answer results when the future value is calculated instead of the present value, and the signs are reversed.

(B) This incorrect answer results when the signs are reversed.

(D) This incorrect answer results when the future value is calculated instead of the present value.

SOLUTION 20

The vapor mass flow rate is

$$\dot{m} = h_m A(\rho_s - \rho_\infty)$$

The vapor density at the water surface is

$$\rho_s = \rho_{\text{sat},60^\circ} = 0.829 \times 10^{-3}\ \text{lbm/ft}^3$$

The vapor density in the free stream is

$$\begin{aligned}\rho_\infty &= \phi\rho_{\text{sat},70^\circ} = (0.50)\left(1.15 \times 10^{-3}\ \frac{\text{lbm}}{\text{ft}^3}\right)\\ &= 0.575 \times 10^{-3}\ \text{lbm/ft}^3\end{aligned}$$

Thus,

$$\begin{aligned}\dot{m} &= \left(0.10\ \frac{\text{ft}}{\text{sec}}\right)(4\ \text{ft}^2)\left(\begin{array}{r}0.829 \times 10^{-3}\ \dfrac{\text{lbm}}{\text{ft}^3}\\ -0.575 \times 10^{-3}\ \dfrac{\text{lbm}}{\text{ft}^3}\end{array}\right)\\ &\quad \times \left(3600\ \frac{\text{sec}}{\text{hr}}\right)\\ &= 0.365\ \text{lbm/hr} \quad (0.37\ \text{lbm/hr})\end{aligned}$$

The answer is (B).

Why Other Options Are Wrong

(A) This incorrect answer results when the area is not included in the calculation.

(B) This incorrect answer results when the relative humidity is neglected and the vapor densities are reversed.

(D) This incorrect answer results when the temperatures of air and water are switched.

SOLUTION 21

The NPSHA is defined as

$$\text{NPSHA} = h_{\text{atm}} + h_{z,S} - h_{f,S} - h_{\text{vp}}$$

Item I will increase the NPSHA because a larger diameter pipe will reduce the fluid velocity. This will reduce the friction loss, h_f.

Item II will *not* increase the NPSHA because throttling the input will increase the fluid velocity. This will increase the friction head loss and decrease the NPSHA.

Item III will increase the NPSHA because pressurizing the supply tank will increase the atmospheric head, h_a.

Item IV will increase the NPSHA because increasing the height increases the static suction head, h_s.

The answer is (D).

Why Other Options Are Wrong

(A) This answer is wrong because item III will also increase the NPSHA.

(B) This answer is wrong because item II will not increase the NPSHA.

(C) This answer is wrong because item II will not increase the NPSHA.

SOLUTION 22

Item I is a pipe fitting classification and is an abbreviation for American National Standard taper pipe thread. NPT is also used as an abbreviation for the following.

American Standard Pipe Tapered Thread
National Pipe Thread
National Pipe Tapered

Item II is a pipe fitting classification and is an abbreviation for the Society of Automotive Engineers.

Item III is a screw thread classification and is an abbreviation for Unified Coarse.

Item IV is the abbreviation for the American Society of Mechanical Engineers.

The answer is (A).

Why Other Options Are Wrong

(B) This answer is incorrect because UNC is a screw thread classification, not a pipe fitting classification.

(C) This answer is incorrect because UNC is a screw thread classification, not a pipe fitting classification.

(D) This answer is incorrect because neither UNC nor ASME is a pipe fitting classification.

SOLUTION 23

The heat transfer rate needed to obtain the required steam production is

$$\begin{aligned}Q &= \dot{m}\Delta h_{\text{water}}\\ &= \dot{m}(h_{\text{steam}} - h_{\text{water}})\end{aligned}$$

The enthalpy of the saturated water entering the drum boiler at 600°F is

$$h_{\text{water}} = 616.7\ \frac{\text{Btu}}{\text{lbm}}$$

The enthalpy of steam at 1900 psia and 95% quality can be determined from the saturated steam tables and the definition of the quality.

$$\begin{aligned} h_{\text{steam}} &= h_f + x h_{fg} \\ &= 660.1\ \frac{\text{Btu}}{\text{lbm}} + (0.95)\left(483.4\ \frac{\text{Btu}}{\text{lbm}}\right) \\ &= 1119.3\ \text{Btu/lbm} \end{aligned}$$

Only 92% of the transferred heat is used, so the total heat transferred from the flue gas must be

$$\begin{aligned} Q_{\text{total}} &= \frac{\dot{m}(h_{\text{steam}} - h_{\text{water}})}{\eta} \\ &= \frac{\left(540{,}000\ \frac{\text{lbm}}{\text{hr}}\right)\left(1119.3\ \frac{\text{Btu}}{\text{lbm}} - 616.7\ \frac{\text{Btu}}{\text{lbm}}\right)}{0.92} \\ &= 295{,}000{,}000\ \text{Btu/hr} \end{aligned}$$

The heat transfer rate, as a function of the temperature difference, is given by

$$Q = UA(T_1 - T_2)$$

Solving for the heat-exchange area,

$$\begin{aligned} A &= \frac{Q}{U(T_1 - T_2)} \\ &= \frac{295{,}000{,}000\ \frac{\text{Btu}}{\text{hr}}}{\left(218.2\ \frac{\text{Btu}}{\text{hr-ft}^2\text{-}^\circ\text{F}}\right)(32^\circ\text{F})} \\ &= 42{,}249\ \text{ft}^2 \end{aligned}$$

The total number of tubes required is

$$\begin{aligned} n_{\text{tubes}} &= \frac{A}{A_{\text{tube}}} \\ &= \frac{42{,}249\ \text{ft}^2}{5.3\ \frac{\text{ft}^2}{\text{tube}}} \\ &= 7972\ \text{tubes} \quad (8000\ \text{tubes}) \end{aligned}$$

The answer is (B).

Why Other Options Are Wrong

(A) This incorrect answer results when the total heat transfer is multiplied by the efficiency instead of divided by it.

(C) This incorrect answer results when the steam quality is omitted when calculating the steam enthalpy.

(D) This incorrect answer results when the temperature difference is neglected in the calculation of the required area.

SOLUTION 24

The suction lift can be calculated using the net positive suction head available (NPSHA) equation

$$\text{NPSHA} = h_{\text{atm}} + h_{z,S} + h_{f,S} - h_{\text{vp}}$$

From a table of water properties, the specific weight of water at 80°F is 62.2 lbf/ft^3. Therefore, the atmospheric head is

$$\begin{aligned} h_{\text{atm}} &= \frac{p}{\gamma} = \frac{\left(14.7\ \frac{\text{lbf}}{\text{in}^2}\right)\left(12\ \frac{\text{in}}{\text{ft}}\right)^2}{62.2\ \frac{\text{lbf}}{\text{ft}^3}} \\ &= 34\ \text{ft} \end{aligned}$$

Also from a table of water properties, the vapor pressure head of water h_{vp} at 80°F is 1.17 ft.

The friction loss head is 5 ft. The static suction head can now be calculated.

$$\begin{aligned} \text{NPSHA} &= h_{\text{atm}} + h_{z,S} + h_{f,S} - h_{\text{vp}} \\ 10\ \text{ft} &= 34\ \text{ft} + h_{z,S} - 5\ \text{ft} - 1.17\ \text{ft} \\ h_{z,S} &= -17.8\ \text{ft} \quad (-18\ \text{ft}) \end{aligned}$$

The negative sign indicates that the pump is above the free level of the supply surface and there is 18 ft of suction lift.

The answer is (B).

Why Other Options Are Wrong

(A) This answer results when the net positive suction head and the suction lift are incorrectly thought to be equal.

(C) This incorrect answer results when the atmospheric head is taken as 43 ft instead of 34 ft.

(D) This incorrect answer results when the friction loss head and vapor pressure head are added to the right side of the NPSHA equation instead of subtracted.

SOLUTION 25

Method 1

A general rule of thumb for quickly calculating the inside pipe diameter is

$$d_i = \sqrt{\frac{Q}{20}} = \sqrt{\frac{1750 \ \frac{\text{gal}}{\text{min}}}{20 \ \frac{\text{gal}}{\text{min-in}^2}}} = 9.35 \text{ in}$$

The smallest available pipe, then, is 12 in.

Method 2

The recommended maximum velocity for water flowing through a city's water pipeline is 2 to 5 ft/sec. Using the flow rate equation, the inner pipe diameter can be found. Using the largest velocity in the flow rate equation, the smallest inner pipe diameter can be found.

$$\begin{aligned} Q &= \text{v}A \\ &= \left(1750 \ \frac{\text{gal}}{\text{min}}\right)\left(\frac{1 \text{ ft}^3}{7.48 \text{ gal}}\right)\left(\frac{1 \text{ min}}{60 \text{ sec}}\right)\left(12 \ \frac{\text{in}}{\text{ft}}\right)^3 \\ &\quad \times \left(5 \ \frac{\text{ft}}{\text{sec}}\right)\left(12 \ \frac{\text{in}}{\text{ft}}\right)\left(\frac{\pi}{4}d^2\right) \\ d &= 11.9 \text{ in} \end{aligned}$$

The smallest schedule-40 pipe with the required inner diameter is 12 in.

The answer is (B).

Why Other Options Are Wrong

(A) This incorrect solution results from dividing by 20 $\text{ft}^3/\text{min-in}^2$ instead of 20 gal/min-in^2.

(C) This incorrect solution results from dividing by 2 gal/min-in^2 instead of 20 gal/min-in^2.

(D) This incorrect solution results from dividing by 20 $\text{ft}^3/\text{min-ft}^2$ instead of 20 gal/min-in^2.

SOLUTION 26

A boiler horsepower is defined as approximately 33,475 Btu/hr, or 9.81 kW, and is the amount of power needed to convert 34.5 lbm/hr of feedwater at 212°F and atmospheric pressure to dry, saturated steam at the same temperature and pressure. Answer IV can be seen to be true from the following calculation.

From the steam tables at 212°F,

$$\begin{aligned} h_{fg} &= 970.3 \text{ Btu/lbm} \\ P &= \dot{m}\Delta h = \left(34.5 \ \frac{\text{lbm}}{\text{hr}}\right)\left(970.3 \ \frac{\text{Btu}}{\text{lbm}}\right) \\ &= 33{,}475 \text{ Btu/hr} \end{aligned}$$

The answer is (C).

Why Other Options Are Wrong

(A) This answer is incorrect because IV is also true.

(B) This answer is incorrect because III is not true, and IV is.

(D) This answer is incorrect because II and III are both false.

SOLUTION 27

Section I of the ASME *Boiler and Pressure Vessel Code* states that the relieving capacities of steam relief safety valves are given by

$$W_{\text{th}} = 51.5Ap$$

The theoretical flow, W_{th}, is in lbm/hr; the nozzle throat area, A, is in square inches; and the pressure, p, for 3% accumulation, is

$$\begin{aligned} p &= 1.03(\text{set pressure}) + 14.7 \ \frac{\text{lbf}}{\text{in}^2} \\ &= (1.03)\left(1400 \ \frac{\text{lbf}}{\text{in}^2}\right) + 14.7 \ \frac{\text{lbf}}{\text{in}^2} \\ &= 1456.7 \text{ lbf/in}^2 \quad [\text{pressure absolute}] \end{aligned}$$

Note that the ASME code says to use for p either the preceding equation or the one following, whichever is greater.

$$p = \text{set pressure} + 2 + 14.7 \ \frac{\text{lbf}}{\text{in}^2}$$

The latter equation, however, will only be greater for set pressures equal to or less than 67 psig.

Substituting into the steam flow equation and solving for A gives

$$\begin{aligned} W_{\text{th}} &= 51.5Ap \\ 200{,}000 \ \frac{\text{lbm}}{\text{hr}} &= \left(51.5 \ \frac{\text{lbm}}{\text{hr-lbf}}\right)A\left(1456.7 \ \frac{\text{lbf}}{\text{in}^2}\right) \\ A &= 2.7 \text{ in}^2 \end{aligned}$$

The answer is (A).

Why Other Options Are Wrong

(B) This incorrect answer results from using an older ASME code giving the relieving capacity as $W_T = 45AP$.

(C) This incorrect answer results from misreading 51.5 as 15.5.

(D) This incorrect answer results from using 140 psig in the calculation instead of 1400 psig.

SOLUTION 28

In the ASME *Boiler and Pressure Vessel Code*, Sec. VIII, Div. 1, the reinforcing area of the shell plus the reinforcing area of the nozzle must equal at least the required reinforcement area.

$$A_s + A_n \geq A_r$$

Because the nozzle abuts the vessel wall, and using a correction factor of 1.0, the required reinforcement area is found using

$$A_r = dt_r$$

d is the finished diameter of the circular opening in the shell, and t_r is the required thickness of the shell based on the circumferential stress.

The required thickness of the shell is found using

$$t_{\text{req}} = \frac{PR}{SE - 0.6P}$$

The allowable stress, S, of SA-106 grade B material from ASME Sec. II is 17.1 ksi, and the efficiency, E, of the joint in a seamless shell is 1.

$$\begin{aligned} t_{\text{req}} &= \frac{PR}{SE - 0.6P} \\ &= \frac{\left(250\ \frac{\text{lbf}}{\text{in}^2}\right)(34\ \text{in})}{\left(17{,}100\ \frac{\text{lbf}}{\text{in}^2}\right)(1.0) - (0.6)\left(250\ \frac{\text{lbf}}{\text{in}^2}\right)} \\ &= 0.501\ \text{in} \end{aligned}$$

Using pipe size tables, the required reinforcement area is found to be

$$A_{\text{req}} = (4.5\ \text{in})(0.501\ \text{in}) = 2.25\ \text{in}^2$$

The area available in the shell is

$$\begin{aligned} A_s &= d(t - t_{\text{req}}) = (4.5\ \text{in})(0.625\ \text{in} - 0.501\ \text{in}) \\ &= 0.558\ \text{in}^2 \\ A_s + A_n &= 0.558\ \text{in}^2 + 0.518\ \text{in}^2 = 1.08\ \text{in}^2 \end{aligned}$$

Since this is less than the required area, the amount of additional reinforcement needed is

$$A_{\text{add}} = 2.25\ \text{in}^2 - 1.08\ \text{in}^2 = 1.17\ \text{in}^2 \quad (1.2\ \text{in}^2)$$

The answer is (B).

Why Other Options Are Wrong

(A) This incorrect answer results when the wrong diameter for the nozzle is used.

(C) This incorrect answer results when an incorrect efficiency of 0.6 is used to calculate t_{req}.

(D) This incorrect answer results when the vessel's diameter is used to calculate t_{req}.

SOLUTION 29

The remaining corrosion allowance (RCA) is found using

$$\text{RCA} = t_{\text{actual}} - t_{\text{min}}$$

ASME code B31.3 says that either of the following equations can be used to determine the minimum thickness, t_{min}.

$$t_{\text{min}} = \frac{Pd}{2(SE + PY)}$$

$$t_{\text{min}} = \frac{Pd}{2SE}$$

Y is a coefficient obtained from a table. The second equation is the one more often used to determine the minimum thickness. Because the stress value is usually so much larger than the internal pressure, the PY term in the denominator of the first equation is frequently negligible.

From schedule-40 pipe tables, the outside diameter of 3 in pipe is 3.5 in. From the ASME *Boiler and Pressure Vessel Code*, Sec. II, the stress value, S, for SA-106 grade B pipe at 600°F is 17.1 ksi. The quality factor, E, for seamless pipe is 1. The minimum thickness is

$$\begin{aligned} t_{\text{min}} &= \frac{Pd}{2SE} \\ &= \frac{\left(250\ \frac{\text{lbf}}{\text{in}^2}\right)(3.5\ \text{in})}{(2)\left(17{,}100\ \frac{\text{lbf}}{\text{in}^2}\right)(1)} \\ &= 0.0256\ \text{in} \end{aligned}$$

The RCA is

$$\text{RCA} = 0.175 \text{ in} - 0.0256 \text{ in} = 0.149 \text{ in} \quad (0.15 \text{ in})$$

The answer is (B).

Why Other Options Are Wrong

(A) This incorrect solution results from neglecting the 2 in the denominator.

(C) This solution results from using a wrong diameter of 3 in, a wrong stress value of 20,000 lbf/in^2, and a factor of safety of 1.67, which is not needed because ASME Sec. II gives the corrected stress value.

(D) This incorrect solution results from using the original pipe thickness for the actual thickness.

SOLUTION 30

Since the viscosity of the oil is temperature dependent, a trial-and-error solution method is required. Design charts from Raimondi and Boyd assume, however, that the oil's viscosity is constant as it passes through the bearing. (These charts can be found in *Mechanical Engineering Design* by Shigley and Mischke.)

Therefore, the average of the oil inlet and exit temperatures is used to determine the viscosity. Begin by assuming the temperature rise will be 60°F. The average oil temperature is

$$T_{\text{ave}} = \frac{120°\text{F} + 180°\text{F}}{2} = 150°\text{F}$$

From a viscosity-temperature chart, the viscosity (μ) of SAE 30 oil at 150°F is 3.6×10^{-6} reyns.

The temperature rise for an oil with this viscosity is found using the equation

$$\Delta T_F = \left(\frac{\left(0.103 \ \frac{\text{in}^2\text{-°F}}{\text{lbf}}\right)P}{1 - \left(\frac{1}{2}\right)\left(\frac{Q_s}{q}\right)}\right)\left(\frac{\left(\frac{r}{c}\right)f}{\frac{Q}{rcnL}}\right)$$

The friction variable $(r/c)f$, the flow rate Q_s/q, and the flow variable $Q/rcnL$ are found using charts where these variables are plotted against the Sommerfeld number for various L/d ratios. The Sommerfeld number is

$$S = \left(\frac{r}{c_r}\right)^2 \left(\frac{\mu n}{P}\right)$$

The load per unit of projected bearing, P, is

$$P = \frac{W}{2rL} = \frac{495 \text{ lbf}}{(2)(0.75 \text{ in})(1.5 \text{ in})} = 220 \text{ lbf/in}^2$$

The L/d ratio is equal to 1, since the bearing length, L, is equal to the diameter.

$$\begin{aligned} S &= \left(\frac{r}{c_r}\right)^2 \left(\frac{\mu n}{P}\right) \\ &= \left(\left(\frac{0.75 \text{ in}}{2 \text{ mils}}\right)\left(1000 \ \frac{\text{mils}}{\text{in}}\right)\right)^2 \\ &\quad (3.6 \times 10^{-6} \text{ reyn})\left(1 \ \frac{\frac{\text{lbf-sec}}{\text{in}^2}}{\text{reyn}}\right) \\ &\quad \times \frac{\times \left(3600 \ \frac{\text{rev}}{\text{min}}\right)\left(\frac{1 \text{ min}}{60 \text{ sec}}\right)}{220 \ \frac{\text{lbf}}{\text{in}^2}} \\ &= 0.14 \end{aligned}$$

From the charts,

$$\left(\frac{r}{c}\right)f = 3.6$$

$$\frac{Q}{rcnL} = 4.3$$

$$\frac{Q_s}{Q} = 0.65$$

Using the earlier equation, the temperature rise is

$$\begin{aligned} \Delta T_F &= \left(\frac{\left(0.103 \ \frac{\text{in}^2\text{-°F}}{\text{lbf}}\right)\left(220 \ \frac{\text{lbf}}{\text{in}^2}\right)}{1 - \left(\frac{1}{2}\right)(0.65)}\right)\left(\frac{3.6}{4.3}\right) \\ &= 28°\text{F} \end{aligned}$$

The average temperature based on this temperature rise is

$$T_{\text{ave}} = T_{\text{in}} + \frac{\Delta T_F}{2} = 120°\text{F} + \frac{28°\text{F}}{2} = 134°\text{F}$$

Now the ordered pair (134°F, 3.6×10^{-6} reyns) can be graphed on the viscosity-temperature chart. Graphing this point and labeling it point A, it can be seen to be below the SAE 30 line, and thus the viscosity used was too low. Choosing a viscosity above the SAE 30 line—say, 6.0×10^{-6} reyns—will give the following temperature rise.

$$S = \left(\left(\frac{0.75\text{ in}}{2\text{ mils}}\right)\left(1000\ \frac{\text{mils}}{\text{in}}\right)\right)^2 \times \frac{(6\times 10^{-6}\text{ reyn})\left(1\ \frac{\frac{\text{lbf-sec}}{\text{in}^2}}{\text{reyn}}\right)\times\left(3600\ \frac{\text{rev}}{\text{min}}\right)\left(\frac{1\text{ min}}{60\text{ sec}}\right)}{220\ \frac{\text{lbf}}{\text{in}^2}} = 0.23$$

From the charts,

$$\left(\frac{r}{c}\right)f = 5.5$$

$$\frac{Q}{rcnL} = 4.05$$

$$\frac{Q_s}{Q} = 0.53$$

The temperature rise is

$$\Delta T_F = \left(\frac{\left(0.103\ \frac{\text{in}^2\text{-°F}}{\text{lbf}}\right)\left(220\ \frac{\text{lbf}}{\text{in}^2}\right)}{1-\left(\frac{1}{2}\right)(0.53)}\right)\left(\frac{5.5}{4.05}\right) = 42°\text{F}$$

The average temperature based on this temperature rise is

$$T_{\text{ave}} = T_{\text{in}} + \frac{\Delta T_F}{2} = 120°\text{F} + \frac{42°\text{F}}{2} = 141°\text{F}$$

Now the ordered pair (141°F, 6.0×10^{-6} reyns) can be graphed on the viscosity-temperature chart and labeled as point B. Draw a line between points A and B and find the intersection of this line with the SAE 30 line. This point gives the correct oil viscosity and average temperature for this bearing under these conditions. The intersection point is (138°F, 4.8×10^{-6} reyns).

The temperature rise is

$$T_{\text{ave}} = \frac{T_{\text{in}} + (T_{\text{in}} + \Delta T)}{2} = T_{\text{in}} + \frac{\Delta T}{2}$$

$$\Delta T = 2(T_{\text{ave}} - T_{\text{in}}) = (2)(138°\text{F} - 120°\text{F}) = 36°\text{F}$$

The answer is (C).

Why Other Options Are Wrong

(A) This incorrect answer results when the factor of two is neglected.

(B) This incorrect answer results when the $(r/c)f$ and $Q/rcnL$ values are switched when calculating T_F for point B.

(D) This incorrect solution results when the temperature T_{ave} is misread from the chart as 148°F.

SOLUTION 31

The quality of the gas leaving the heater can be calculated from the formula

$$x = \frac{h_{2,\text{out}} - h_f}{h_{fg}}$$

h_f and h_{fg} can be found from R-12 tables at 120 psia, and equal 29.53 Btu/lbm and 58.52 Btu/lbm, respectively. The enthalpy of the gas leaving the heater, $h_{2,\text{out}}$, must be calculated.

The energy gained on the liquid side of the heater must equal the energy lost on the gas side. An energy balance on the heater gives

$$\dot{m}_1(h_{1,\text{in}} - h_{1,\text{out}}) = \dot{m}_2(h_{2,\text{out}} - h_{2,\text{in}})$$

$$h_{2,\text{out}} = h_{2,\text{in}} - \left(\frac{\dot{m}_1}{\dot{m}_2}\right)(h_{1,\text{out}} - h_{1,\text{in}})$$

The enthalpy of the saturated liquid leaving the heater, $h_{1,\text{out}}$, and the enthalpy of the superheated gas entering the heater, $h_{2,\text{in}}$, can be found from the given information and R-12 tables, and are equal to 26.49 Btu/lbm and 89.13 Btu/lbm, respectively.

The enthalpy of the saturated liquid entering the heater, $h_{1,\text{in}}$, is equal to its enthalpy entering the pump plus the enthalpy added to it by the pump. Because the liquid is known to enter the pump at 0°F and a quality of zero, its enthalpy can be found from R-12 tables and is equal to 8.25 Btu/lbm.

The enthalpy added by the pump can be calculated from the drop in pressure and the specific volume of the saturated liquid. The discharge pressure, p_d, is known to be 100 psia. The suction pressure, p_s, for the liquid at a temperature of 0°F can be found from R-12 tables and is equal to 23.87 psia. The specific volume can also be found from R-12 tables. At an estimated average pressure of 60 psia, the specific volume equals 0.0117 ft^3/lbm.

Neglecting velocity and potential heads, the enthalpy of the liquid entering the heater is

$$\begin{aligned} h_{1,\text{in}} &= h_{\text{pump,out}} + h_{\text{pump,added}} \\ &= h_{\text{pump,in}} + v(p_d - p_s) \\ &= 8.25\ \frac{\text{Btu}}{\text{lbm}} + \left(0.0117\ \frac{\text{ft}^3}{\text{lbm}}\right) \\ &\quad \times \left(100\ \frac{\text{lbf}}{\text{in}^2} - 23.87\ \frac{\text{lbf}}{\text{in}^2}\right)\left(\frac{1\ \text{Btu}}{778\ \text{ft-lbf}}\right) \\ &\quad \times \left(12\ \frac{\text{in}}{\text{ft}}\right)^2 \\ &= 8.41\ \text{Btu/lbm} \end{aligned}$$

The enthalpy of the gas exiting the heater can now be calculated.

$$\begin{aligned} h_{2,\text{out}} &= h_{2,\text{in}} - \left(\frac{\dot{m}_1}{\dot{m}_2}\right)\left(h_{1,\text{out}} - h_{1,\text{in}}\right) \\ &= 89.13\ \frac{\text{Btu}}{\text{lbm}} - \left(\frac{1\ \frac{\text{lbm}}{\text{sec}}}{0.5\ \frac{\text{lbm}}{\text{sec}}}\right) \\ &\quad \times \left(26.49\ \frac{\text{Btu}}{\text{lbm}} - 8.41\ \frac{\text{Btu}}{\text{lbm}}\right) \\ &= 53.0\ \text{Btu/lbm} \end{aligned}$$

The quality of the gas exiting the heater is

$$\begin{aligned} x &= \frac{h_{2,\text{out}} - h_f}{h_{fg}} \\ &= \frac{53.0\ \frac{\text{Btu}}{\text{lbm}} - 29.53\ \frac{\text{Btu}}{\text{lbm}}}{58.52\ \frac{\text{Btu}}{\text{lbm}}} \\ &= 0.401 \quad (40\%) \end{aligned}$$

The answer is (B).

Why Other Options Are Wrong

(A) This incorrect answer results when the R-12 tables are misread and h_g is used instead of h_{fg}.

(C) This incorrect answer results when the conversion from Btu to ft-lbf is neglected.

(D) This incorrect answer results when the mass flow rates are switched.

SOLUTION 32

The flow rate of the cooling water in gallons per minute can be approximated by

$$Q = \frac{H(\text{BHP})}{\left(500\ \frac{\text{Btu-hr-min}}{{}^\circ\text{F-gal}}\right)\Delta T}$$

The heat dissipation, H, will vary for different engines, and the value supplied by the manufacturer should be used. If it is unavailable, then for engines with water-cooled exhaust manifolds 600 Btu-hr/bhp is a safe value to use.

Usually manufacturers recommend that the temperature difference across the engine, ΔT, not exceed 15°F, with 10°F being preferable. Using 10°F, the flow rate is

$$Q = \frac{\left(600\ \frac{\text{Btu-hr}}{\text{bhp}}\right)(1100\ \text{bhp})}{\left(500\ \frac{\text{Btu-hr-min}}{{}^\circ\text{F-gal}}\right)(10^\circ\text{F})} = 132\ \text{gal/min}$$

The answer is (B).

Why Other Options Are Wrong

(A) This is the number of cubic feet per minute needed, not gallons per minute.

(C) This incorrect solution results from using 2200 Btu-hr/bhp for H in the calculation of the required flow rate. This is the value that should be used when calculating the engine-jacket water requirement, not the lube oil's cooling water requirement.

(D) This incorrect solution results from not dividing by 10°F when calculating the flow rate.

SOLUTION 33

The steam quality is found using the equation

$$h = h_f + xh_{fg}$$

From the steam tables at 1800 psia,

$$\begin{aligned} h_f &= 648.3\ \text{Btu/lbm} \\ h_{fg} &= 502.1\ \text{Btu/lbm} \end{aligned}$$

The total enthalpy in the steam, h, must be found. This needs to be equal to 800 Btu/lbm plus the initial enthalpy in the feedwater. At 280°F the enthalpy of the feedwater is

$$h_{\text{feed}} = 249.18\ \text{Btu/lbm}$$

The total enthalpy, then, is

$$\begin{aligned} h_{\text{total}} &= h_{\text{added}} + h_{\text{feed}} \\ &= 800\ \frac{\text{Btu}}{\text{lbm}} + 249.18\ \frac{\text{Btu}}{\text{lbm}} \\ &= 1049\ \text{Btu/lbm} \end{aligned}$$

The quality can now be found.

$$\begin{aligned} h &= h_f + xh_{fg} \\ 1049\ \frac{\text{Btu}}{\text{lbm}} &= 648.3\ \frac{\text{Btu}}{\text{lbm}} + x\left(502.1\ \frac{\text{Btu}}{\text{lbm}}\right) \\ x &= 0.798 \quad (80\%) \end{aligned}$$

The answer is (C).

Why Other Options Are Wrong

(A) This incorrect answer results when the feedwater enthalpy is subtracted from, instead of added to, the 800 Btu/lbm.

(B) This incorrect answer results when 800 Btu/lbm is used for the total energy.

(C) This incorrect answer results when 1094 is entered into the calculator instead of 1049.

SOLUTION 34

Statement I is false because, as the hint suggests, isothermal compression requires less power.

Statement II is also false. An isentropic process is both adiabatic and reversible, but an adiabatic process need not be reversible. So although all isentropic processes are adiabatic, not all adiabatic processes are isentropic.

Statement III is true. The work of an isentropic air compressor is

$$W = \dot{m}(h_2 - h_1) = \dot{m}c_p(T_2 - T_1)$$

For air, the specific heat is 0.244 Btu/lbm-°R and the ratio of specific heats is 1.4. The final temperature can be calculated from the initial temperature by the equation

$$\begin{aligned} T_2 &= T_1\left(\frac{p_2}{p_1}\right)^{\frac{k-1}{k}} \\ &= (530°\text{R})\left(\frac{200\ \frac{\text{lbf}}{\text{in}^2}}{14.7\ \frac{\text{lbf}}{\text{in}^2}}\right)^{\frac{1.4-1}{1.4}} \\ &= 1120°\text{R} \end{aligned}$$

The work can now be calculated.

$$\begin{aligned} W &= \dot{m}c_p(T_2 - T_1) \\ &= \left(1\ \frac{\text{lbm}}{\text{sec}}\right)\left(3600\ \frac{\text{sec}}{\text{hr}}\right)\left(0.24\ \frac{\text{Btu}}{\text{lbm-°R}}\right) \\ &\quad \times (1120\ °\text{R} - 530°\text{R})\left(\frac{1\ \text{hp}}{2545\ \frac{\text{Btu}}{\text{hr}}}\right) \\ &= 200\ \text{hp} \end{aligned}$$

Thus, only statement III is true.

The answer is (C).

Why Other Options Are Wrong

(A) This answer is incorrect because statement I is false.

(B) This answer is incorrect because statement II is false.

(D) This answer is incorrect because statement I is false.

SOLUTION 35

From the ASME *Boiler and Pressure Vessel Code*, Sec. VIII, Div. 1, the minimum thickness is calculated using the equation

$$\begin{aligned} t_{\min} &= d\sqrt{\frac{Cp}{SE}} \\ &= (3\ \text{ft})\left(12\ \frac{\text{in}}{\text{ft}}\right)\sqrt{\frac{(0.20)\left(150\ \frac{\text{lbf}}{\text{in}^2}\right)}{\left(17{,}500\ \frac{\text{lbf}}{\text{in}^2}\right)(1)}} \\ &= 1.5\ \text{in} \end{aligned}$$

The thickness of the head can then be calculated using the equation

$$\begin{aligned} t_{\text{head}} &= t_{\min} + (\text{CR})L \\ &= 1.5\ \text{in} + \left(8\ \frac{\text{mil}}{\text{yr}}\right)\left(\frac{1\ \text{in}}{1000\ \text{mil}}\right)(30\ \text{yr}) \\ &= 1.74\ \text{in} \quad (1.7\ \text{in}) \end{aligned}$$

The answer is (C).

Why Other Options Are Wrong

(A) The minimum design thickness must be greater than just the corrosion amount of 0.24 in over 30 years or the head will not be able to withstand the pressure.

(B) This incorrect answer results when the diameter is not converted to inches.

(D) This incorrect answer results when the attachment factor is ignored.

SOLUTION 36

From the ASME *Boiler and Pressure Vessel Code*, Sec. VIII, Div. 1, the thickness is calculated using the equation

$$t = d\sqrt{\frac{Cp}{SE}}$$

Since a blind flange is normally seamless, the joint efficiency, E, is equal to 1.

$$\begin{aligned} t &= (2 \text{ ft})\left(12 \ \frac{\text{in}}{\text{ft}}\right)\sqrt{\frac{(0.17)\left(200 \ \frac{\text{lbf}}{\text{in}^2}\right)}{\left(13{,}000 \ \frac{\text{lbf}}{\text{in}^2}\right)(1)}} \\ &= 1.2 \text{ in} \end{aligned}$$

The answer is (B).

Why Other Options Are Wrong

(A) This incorrect answer results when the diameter is not converted to inches.

(C) This incorrect answer results when the attachment factor is ignored.

(D) This incorrect answer results when the diameter is not converted to inches and the design pressure and allowable stress are interchanged.

SOLUTION 37

API 682 states that the injection flow rate for piping plan 12 can be determined using the equation

$$Q_{\text{inj}} = \frac{14.35P}{(\text{SG})(dT)c_p}$$

When using this equation, this standard states that a design factor of least two should be applied to the flow rate.

To make the injection rate equation dimensionally correct, the constant must have the following units.

$$Q_{\text{inj}} = \left(14.35 \ \frac{\text{L·s}}{\text{kg·min}}\right)\left(\frac{P}{(\text{SG})(dT)c_p}\right)$$

To use this equation, the heat generation at the seal forces, P, must be in kilowatts; the differential temperature, dT, between the pump and the desired seal chamber temperature must be in degrees Celsius; and the specific heat, c_p, of the injected fluid must be in kJ/kg·°C.

API 682 states that for good performance the temperature rise should be between 2.8°C and 5.6°C, inclusive. Therefore, the minimum allowable injection flow rate is the one that gives a value of 5.6°C for dT.

For water,

$$\begin{aligned} \text{SG} &= 1.0 \\ c_p &= 4.18 \text{ kJ/kg·°C} \end{aligned}$$

Before allowing for the design factor, the minimum allowable injection flow rate is

$$\begin{aligned} Q_{\text{inj}} &= \left(14.35 \ \frac{\text{L·s}}{\text{kg·min}}\right)\left(\frac{(1.5 \text{ kW})\left(1 \ \frac{\frac{\text{kJ}}{\text{s}}}{\text{kW}}\right)}{(1.0)(5.6\text{°C})\left(4.18 \ \frac{\text{kJ}}{\text{kg·°C}}\right)}\right) \\ &= 0.92 \text{ L/min} \end{aligned}$$

The design factor of two makes the minimum allowable injection flow rate approximately 1.8 L/min.

The answer is (B).

Why Other Options Are Wrong

(A) This incorrect answer results when the design factor of two is neglected.

(C) This incorrect answer results when a temperature rise of 2.8°C is used instead of 5.6°C.

(D) This incorrect answer results when the numerical value for the specific heat in Btu/lbm-°F is mistakenly used.

SOLUTION 38

The volume of air is reduced to a 1/16 of its original volume, and the mass of gas remains the same, so the density of the gas is increased by a factor of 16. The ratio of specific heat, k, for air is 1.4, and the initial pressure p_1 is atmospheric at 14.7 psi.

For isentropic compression,

$$\frac{p_1}{\rho_1^k} = \frac{p_2}{\rho_2^k}$$

The subscripts 1 and 2 represent the initial and final states, respectively. Solving for p_2 gives

$$\begin{aligned} p_2 &= p_1\left(\frac{\rho_2}{\rho_1}\right)^k \\ &= \left(14.7\ \frac{\text{lbf}}{\text{in}^2}\right)\left(\frac{16\rho_1}{\rho_1}\right)^{1.4} \\ &= 713\ \text{lbf/in}^2 \quad (710\ \text{psi}) \end{aligned}$$

The answer is (D).

Why Other Options Are Wrong

(A) This incorrect answer results when the 1.4th root of 16 is taken instead of the 1.4th power.

(B) This incorrect answer results when the initial pressure (atmospheric, 14.7 psi) is multiplied by the compression ratio instead of divided by it.

(C) This incorrect answer results when an incorrect ratio of specific heats of 1.3 is used.

SOLUTION 39

The Mach number can be determined using the isentropic flow table if the isentropic flow factor p/p_0 is known. The stagnation pressure, or total pressure, p_0, is given as 500 kPa.

If $p_{\text{reservoir}}/p_{\text{ambient}}$ is less than a critical value, then the exit static pressure will be the condition that determines the speed of the flow. The exit static pressure is nearly equal to the ambient pressure.

$$p = p_e \approx p_{\text{ambient}} = 101.3\ \text{kPa}$$

This is because large pressure differences cannot occur in subsonic flow over small distances. Any large pressure difference that occurs between a nozzle exit and a point slightly outside the exit will require a shock to maintain the pressure discontinuity, and this can only occur in supersonic flow.

$$\frac{p}{p_0} = \frac{101.3\ \text{kPa}}{500\ \text{kPa}} = 0.2026$$

In the isentropic flow table, a value for p/p_0 of 0.2026 corresponds to a value for M of 1.7.

The answer is (B).

Why Other Options Are Wrong

(A) This incorrect answer results when the value 0.2026 is found in the M_y column of the isentropic flow table.

(C) This incorrect answer results when the value 0.2026 is found in the ρ/ρ_0 column of the isentropic flow table.

(D) This incorrect answer results when the value 0.2026 is found in the T/T_0 column of the isentropic flow table.

SOLUTION 40

According to ASME Standard B31.1, for a maximum operating temperature of 500°F, the maximum pressure must be temperature derated by 0.65. Because the severity of service is normal (A), the severity of service derating factor is equal to 1.

$$\begin{aligned} P_{\text{max}} &= (0.65)\left(1000\ \frac{\text{lbf}}{\text{in}^2}\right) \\ &= 650\ \text{lbf/in}^2 \quad (650\ \text{psi}) \end{aligned}$$

The answer is (A).

Why Other Options Are Wrong

(B) This incorrect answer results when the derating factor at the average temperature, 300°F, is used.

(C) This incorrect answer results when the derating factor for C-1010 steel is used.

(D) This incorrect answer results when no derating factor is used.

SOLUTION 41

Find the flow diameter (inner diameter) of the tubing. The recommended flow velocity for a pressure line is 25 ft/sec. (For a return line it is 10 ft/sec, and for a suction line it is 4 ft/sec.)

$$\begin{aligned} Q &= \text{v}A = \text{v}\left(\frac{\pi d^2}{4}\right) \\ d &= \sqrt{\frac{4Q}{\pi \text{v}}} \\ &= \sqrt{\frac{(4)\left(\dfrac{15\ \dfrac{\text{gal}}{\text{min}}}{\left(7.48\ \dfrac{\text{gal}}{\text{ft}^3}\right)\left(60\ \dfrac{\text{sec}}{\text{min}}\right)}\right)}{\pi\left(25\ \dfrac{\text{ft}}{\text{sec}}\right)}}\left(12\ \frac{\text{in}}{\text{ft}}\right) \\ &= 0.4951\ \text{in} \end{aligned}$$

At a normal (A) severity of service, the safety factor is 4 and the severity of service derating factor is 1; for C-1010 steel at 200°F, the temperature derating factor is also 1. (For severe (B) severity of service, the safety factor is 6 and the derating factor is 0.67; for hazardous (C) severity of service, the safety factor is 8 and the derating factor is 0.5.)

Using Parker Hannifin's pressure ratings table with a maximum pressure of 3000 psi, the following tubing sizes have inner diameters of at least 0.5 in.

0.750×0.072

0.750×0.083

0.750×0.095

0.750×0.109

The 0.750×0.072 tubing, however, has a design pressure of only 2650 psi, so it can't be used. The remaining tubes have design pressures above 3000 psi so they can be used. The lightest tubing is the one with the smallest wall thickness, 0.750×0.083.

The answer is (D).

Why Other Options Are Wrong

(A) This incorrect answer results when an incorrect velocity of 100 ft/sec is used.

(B) This incorrect answer results when 0.5 in is taken as the outer diameter, not the inner.

(C) This incorrect answer results when the first tubing in the pressure ratings table with an inner diameter greater than 0.5 in is selected, but the design pressure is less than 3000 psi.

SOLUTION 42

From a table of pipe dimensions for schedule-40 pipe,

$$A_{\text{A}} = A_{\text{B}} = \frac{0.3039\ \text{in}^2}{\left(12\ \frac{\text{in}}{\text{ft}}\right)^2} = 0.002110\ \text{ft}^2$$

$$A_{\text{C}} = \frac{0.5333\ \text{in}^2}{\left(12\ \frac{\text{in}}{\text{ft}}\right)^2} = 0.003703\ \text{ft}^2$$

Find the water velocity at location C.

$$\begin{aligned} Q_{\text{C}} &= Q_{\text{A}} + Q_{\text{B}} \\ \text{v}_{\text{C}} A_{\text{C}} &= \text{v}_{\text{A}} A_{\text{A}} + \text{v}_{\text{B}} A_{\text{B}} \\ \text{v}_{\text{C}} &= \frac{\text{v}_{\text{A}} A_{\text{A}} + \text{v}_{\text{B}} A_{\text{B}}}{A_{\text{C}}} \\ &= \frac{\left(12\ \frac{\text{ft}}{\text{sec}}\right)(0.002110\ \text{ft}^2) + \left(25\ \frac{\text{ft}}{\text{sec}}\right)(0.002110\ \text{ft}^2)}{0.003703\ \text{ft}^2} \\ &= 21.08\ \text{ft/sec} \end{aligned}$$

Draw a free-body diagram of the pipe at location C.

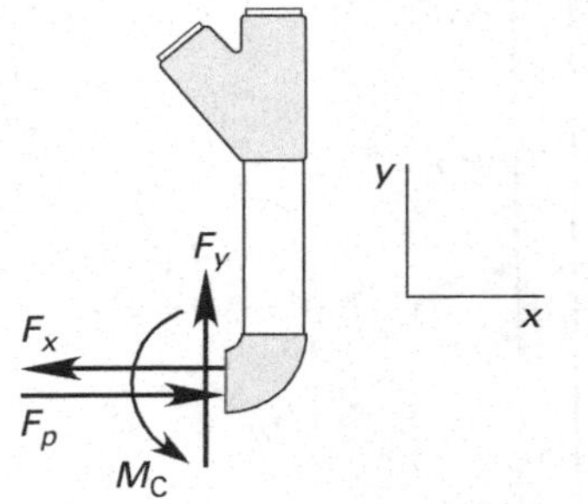

The steady-flow equation in the x-direction is

$$\sum F_x = \frac{dm_{\text{A}}}{dt}\text{v}_{\text{A},x} + \frac{dm_{\text{B}}}{dt}\text{v}_{\text{B},x} - \frac{dm_{\text{C}}}{dt}\text{v}_{\text{C},x}$$

The force due to the static pressure at location C is

$$\begin{aligned} F_p &= pA_{\text{C}} = \left(40\ \frac{\text{lbf}}{\text{in}^2}\right)(0.5333\ \text{in}^2) \\ &= 21.33\ \text{lbf} \end{aligned}$$

Summing forces in the x-direction at location C gives

$$\begin{aligned} F_p - F_x &= \frac{dm_{\text{A}}}{dt}\text{v}_{\text{A},x} + \frac{dm_{\text{B}}}{dt}\text{v}_{\text{B},x} - \frac{dm_{\text{C}}}{dt}\text{v}_{\text{C},x} \\ &= \left(\frac{\rho \text{v} A}{g_c}\right)_{\text{A}}\text{v}_{\text{A},x} + \left(\frac{\rho \text{v} A}{g_c}\right)_{\text{B}}\text{v}_{\text{B},x} \\ &\quad - \left(\frac{\rho \text{v} A}{g_c}\right)_{\text{C}}\text{v}_{\text{C},x} \end{aligned}$$

$$F_x = F_p - \left(\frac{\rho}{g_c}\right) \times \begin{pmatrix} (\mathrm{v}A)_A \mathrm{v}_{A,x} + (\mathrm{v}A)_B \mathrm{v}_{B,x} \\ -(\mathrm{v}A)_C \mathrm{v}_{C,x} \end{pmatrix}$$

$$= 21.33 \text{ lbf} - \left(\frac{62.4 \ \frac{\text{lbm}}{\text{ft}^3}}{32.2 \ \frac{\text{lbm-ft}}{\text{lbf-sec}^2}}\right) \times \begin{pmatrix} \left(12 \ \frac{\text{ft}}{\text{sec}}\right)(0.002110 \text{ ft}^2) \\ \times\left(\left(-12 \ \frac{\text{ft}}{\text{sec}}\right)\sin 30^\circ\right) \\ +\left(25 \ \frac{\text{ft}}{\text{sec}}\right)(0.002110 \text{ ft}^2) \\ \times\left(0 \ \frac{\text{ft}}{\text{sec}}\right) \\ -\left(21.08 \ \frac{\text{ft}}{\text{sec}}\right)(0.003703 \text{ ft}^2) \\ \times\left(21.08 \ \frac{\text{ft}}{\text{sec}}\right) \end{pmatrix}$$

$$= 24.81 \text{ lbf} \quad (25 \text{ lbf})$$

The answer is (D).

Why Other Options Are Wrong

(A) This incorrect answer results when the static pressure is not included in the force calculation.

(B) This incorrect answer results when an incorrect pipe area at A and B is used to compute the static pressure force.

(C) This incorrect answer is the force due to the static pressure alone.

SOLUTION 43

A pressure wave is created upstream of a valve whenever it is closed in a pipe. This wave travels at the speed of sound and is known as water hammer.

At 60°F, the density of water, ρ, is 62.4 lbm/ft^3, and the speed of sound, c, in water is 4800 ft/sec. The pressure change is

$$\Delta p = \frac{n\rho c \Delta \mathrm{v}}{g_c}$$

$$= \frac{(1.1)\left(62.4 \ \frac{\text{lbm}}{\text{ft}^3}\right)\left(4800 \ \frac{\text{ft}}{\text{sec}}\right)\left(3 \ \frac{\text{ft}}{\text{sec}}\right)}{\left(32.2 \ \frac{\text{lbm-ft}}{\text{lbf-sec}^2}\right)\left(12 \ \frac{\text{in}}{\text{ft}}\right)^2}$$

$$= 213.2 \text{ lbf/in}^2$$

The maximum pressure is

$$p_{\text{max}} = p_0 + \Delta p = 60 \ \frac{\text{lbf}}{\text{in}^2} + 213.2 \ \frac{\text{lbf}}{\text{in}^2}$$

$$= 273.2 \text{ lbf/in}^2 \quad (270 \text{ psi})$$

The answer is (D).

Why Other Options Are Wrong

(A) This incorrect answer results when the flow rate is used instead of the velocity.

(B) This incorrect answer results when the speed of sound in meters per second is used instead of feet per second.

(C) This incorrect answer is the change in pressure alone and with no safety factor included.

SOLUTION 44

From the graph, the pump efficiency at 400 gpm is approximately 60%, and the amount of head gained is approximately 590 ft. The amount of pump head needed at the shaft is

$$h_{\text{shaft}} = \frac{h_{\text{gained}}}{\eta} = \frac{590 \text{ ft}}{0.6} = 983 \text{ ft} \quad (980 \text{ ft})$$

The answer is (B).

Why Other Options Are Wrong

(A) This incorrect answer results when the maximum pump efficiency (approximately 70%) is used instead of the pump efficiency at the operating point.

(C) This incorrect answer results when the head value used is taken at the point where the vertical 400 gpm line crosses the efficiency curve (approximately 650 ft).

(D) This incorrect answer results when an efficiency of 40% (that is, 100% – 60%) is used instead of 60%.

SOLUTION 45

The heat flow per unit mass from the condenser is

$$Q_L = h_2 - h_3$$

The enthalpy at state 3, h_3, is given as 94 Btu/lbm. The enthalpy at state 2 is

$$h_2 = h_f + x_2 h_{fg}$$

The enthalpies of saturated liquid at 2 psi are

$$h_f = 94 \text{ Btu/lbm}$$
$$h_{fg} = 1021.7 \text{ Btu/lbm}$$

The steam quality at state 2 is

$$x_2 = \frac{s_2 - s_f}{s_{fg}}$$

The turbine expands the steam isentropically, so

$$s_2 = s_1 = 1.6112 \text{ Btu/lbm-°R}$$

The entropies of saturated liquid at 2 psi are

$$s_f = 0.17499 \text{ Btu/lbm-°R}$$
$$s_{fg} = 1.74444 \text{ Btu/lbm-°R}$$

Therefore, the steam quality at state 2 is

$$\begin{aligned} x_2 &= \frac{s_2 - s_f}{s_{fg}} \\ &= \frac{1.6112 \ \frac{\text{Btu}}{\text{lbm-°R}} - 0.17499 \ \frac{\text{Btu}}{\text{lbm-°R}}}{1.74444 \ \frac{\text{Btu}}{\text{lbm-°R}}} \\ &= 0.8233 \end{aligned}$$

The enthalpy at state 2 is

$$\begin{aligned} h_2 &= h_f + x_2 h_{fg} \\ &= 94 \ \frac{\text{Btu}}{\text{lbm}} + (0.8233)\left(1021.7 \ \frac{\text{Btu}}{\text{lbm}}\right) \\ &= 935.2 \text{ Btu/lbm} \end{aligned}$$

The heat flow per unit mass from the condenser is

$$\begin{aligned} Q_L &= h_2 - h_3 \\ &= 935.2 \ \frac{\text{Btu}}{\text{lbm}} - 94 \ \frac{\text{Btu}}{\text{lbm}} \\ &= 841.2 \text{ Btu/lbm} \quad (840 \text{ Btu/lbm}) \end{aligned}$$

The answer is (B).

Why Other Options Are Wrong

(A) This incorrect answer is the enthalpy at state 3.

(C) This incorrect answer results when a steam quality of 100% is used.

(D) This incorrect answer results when the enthalpy at state 1 is used in the final calculation instead of the enthalpy at state 2.

SOLUTION 46

Statements I and III are true.

Gas turbine power cycles typically operate at higher temperatures than steam power cycles, so statement II is false. The practical upper limit for the efficiency of a combined power cycle is considered to be approximately 70%, so statement IV is false. (A combined cycle power plant in Bouchain, France, developed by GE and EDF, achieved a 62.22% efficiency rating on April 28, 2016.)

The answer is (A).

Why Other Options Are Wrong

(B) This answer is incorrect because statements II and IV are false.

(C) This answer is incorrect because statement II is false.

(D) This answer is incorrect because statement IV is false.

2 Hydraulic and Fluid Applications

PROBLEM 1

A 150 hp reduced crude pump in a refinery pumps 1700 gpm. The brake horsepower needed to pump 2500 gpm is most nearly

(A) 150 hp

(B) 220 hp

(C) 320 hp

(D) 480 hp

Hint: The relationship between the brake horsepower of a pump and its flow rate is nonlinear.

PROBLEM 2

A water system with a capacity of 100 gpm has a total head of 240 ft at the suction line of a single-stage, two-pole, centrifugal pump. The maximum attainable efficiency for the pump is most nearly

(A) 29%

(B) 40%

(C) 52%

(D) 70%

Hint: Calculate the specific speed of the pump.

PROBLEM 3

A centrifugal pump at an oil pipeline pumping station pumps 70,000 barrels/day of crude oil (specific gravity 0.86, kinematic viscosity 250×10^{-6} ft^2/sec) through a 12 in schedule-40 steel pipe to a booster pump inside a refinery 10 mi away. The pump develops 956 ft of head. The motor torque supplied to the pump shaft running at 1750 rpm is 1500 lbf-ft. The pump's efficiency is most nearly

(A) 16%

(B) 67%

(C) 85%

(D) 98%

Hint: Calculate the water horsepower of the pump.

PROBLEM 4

The feedwater piping system of a boiler is designed for an operating pressure of 2200 psi and a temperature of 280°F at the outlet. The friction pressure drop in the piping system under the design conditions is 33 psi. The design flow rate is 1.4×10^6 lbm/hr. The velocity pressure differentials and the static pressure drops are negligible; the lowest mass flow rate is 740,000 lbm/hr. The lowest pressure drop in the feedwater piping is most nearly

(A) 9.2 lbf/in^2

(B) 17 lbf/in^2

(C) 33 lbf/in^2

(D) 120 lbf/in^2

Hint: Assume the friction pressure drop is proportional to the square of the velocity.

PROBLEM 5

A pipeline carries crude oil (specific gravity 0.86, kinematic viscosity 250×10^{-6} ft^2/sec). A 250 hp pump at the pipeline pumping station generates 450 ft of head. The increase in pressure at the pump discharge is most nearly

(A) 170 psi

(B) 200 psi

(C) 2000 psi

(D) 24,000 psi

Hint: Equate feet of head to static pressure.

PROBLEM 6

A Carnot engine operates with a minimum temperature of 80°F, a maximum temperature of 1200°F, and a maximum pressure of 1300 psia. Assume that air behaves as an ideal gas. If 0.2 lbm of air is the working fluid and 5 Btu of heat is added to the air, the maximum cylinder volume is most nearly

(A) 0.10 ft^3

(B) 1.5 ft^3

(C) 2.0 ft^3

(D) 2.7 ft^3

Hint: Begin at the state where the pressure is a maximum.

PROBLEM 7

An air compressor operates at steady state. Air enters through a 1.0 ft^2 opening at a velocity of 20 ft/sec, a temperature of 77°F, and a pressure of 14.7 psia, and exits at a velocity of 6 ft/sec, a temperature of 360°F, and a pressure of 100 psia. The power input to the compressor is 150 hp. Assuming ideal air properties, the heat input to the compressor is most nearly

(A) −48,000 Btu/hr

(B) −19,000 Btu/hr

(C) −18,000 Btu/hr

(D) 0 Btu/hr

Hint: Neglect potential energy changes.

PROBLEM 8

The *T-s* diagram shown represents the air-standard power cycle of an internal combustion engine.

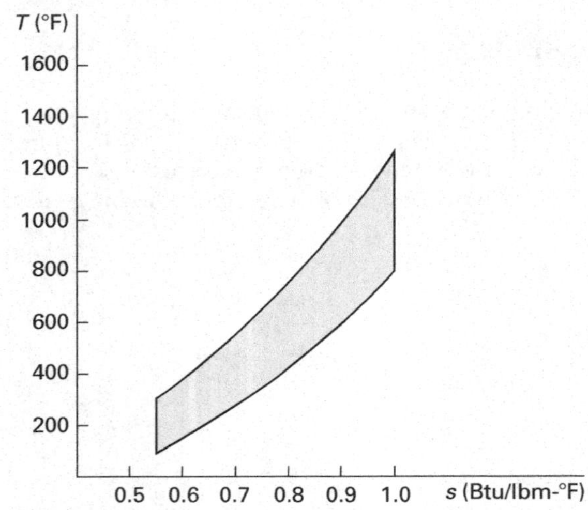

The thermal efficiency of this engine is most nearly

(A) 26%

(B) 36%

(C) 67%

(D) 92%

Hint: Internal combustion engines are evaluated using a particular power cycle.

PROBLEM 9

Which of the following statements about the thermal efficiency of reversible heat engines are true?

I. The higher the temperature at which heat is added, the higher the efficiency.

II. The maximum thermal efficiency is achieved using a Carnot cycle.

III. The lower the temperature at which heat is rejected, the higher the efficiency.

IV. Different reversible engines operating between the same temperature limits have different thermal efficiencies.

(A) I and II

(B) I and IV

(C) II and III

(D) I, II, and III

Hint: This is a nonquantitative question. Start with the index of a thermodynamics book.

PROBLEM 10

A pump rated for 350 gpm and 200 ft of head is required for use as a refinery boiler feedwater pump that will run continuously year-round. The manufacturer's performance curves for a particular pump are shown in *Illustration for Problem 10 and Problem 11.* If the pump is oversized by 20%, the increased yearly electrical cost at $0.07/kW-hr for the oversized pump is most nearly

(A) $0

(B) $915

(C) $2750

(D) $5490

Hint: Be sure to increase both the flow rate and head by 20%, then treat the oversize requirement as a new rating and design accordingly.

Illustration for Problem 10 and Problem 11

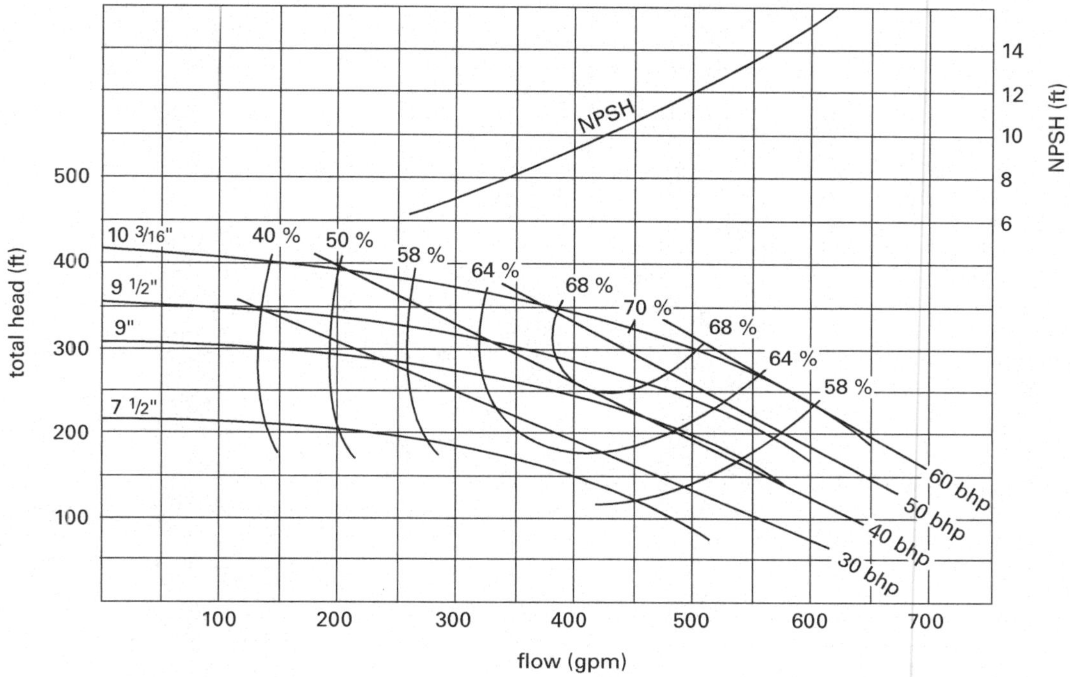

Reprinted/adapted with permission from Flowserve.

PROBLEM 11

A refinery boiler feedwater pump operating at 3550 rpm has a 9.5 in impeller and performance curves as shown in *Illustration for Problem 10 and Problem 11.* If the pump needs to achieve a rating of 550 gpm and 350 ft, the speed at which it should be run is most nearly

(A) 3070 rpm

(B) 4150 rpm

(C) 4740 rpm

(D) 5140 rpm

Hint: Use trial and error.

PROBLEM 12

The maximum horsepower that can be transmitted by a 2.5 in diameter, solid steel cooling tower fan shaft running at 200 rpm is most nearly

(A) 80 hp

(B) 150 hp

(C) 190 hp

(D) 290 hp

Hint: Use a failure criterion.

PROBLEM 13

A crude overhead product pump operating at 3550 rpm has the characteristics shown.

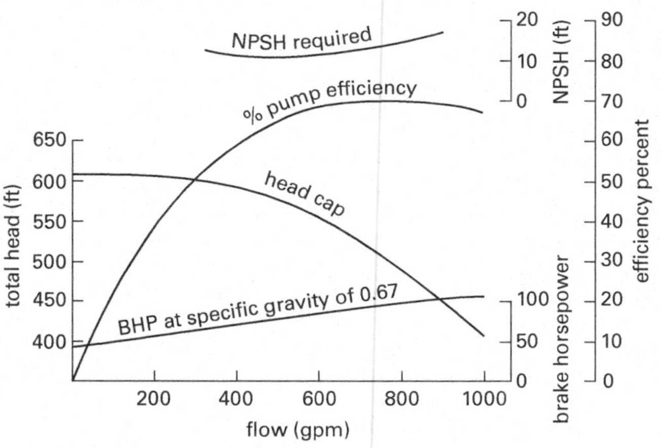

It is connected to a system requiring 200 ft of head when handling 500 gpm of crude oil. The flow rate in the system when the pump is running at 3550 rpm is most nearly

(A) 220 gpm

(B) 300 gpm

(C) 800 gpm

(D) 1000 gpm

Hint: Draw the system curve.

PROBLEM 14

The turbine shown has an output of 10 MW and is 90% efficient. Saturated steam at 1000 psia flows from the boiler through the turbine and exhausts to a condenser at 2 psia. The condensate is subcooled 6°F in the condenser. The condensate flow is 100,000 lbm/hr.

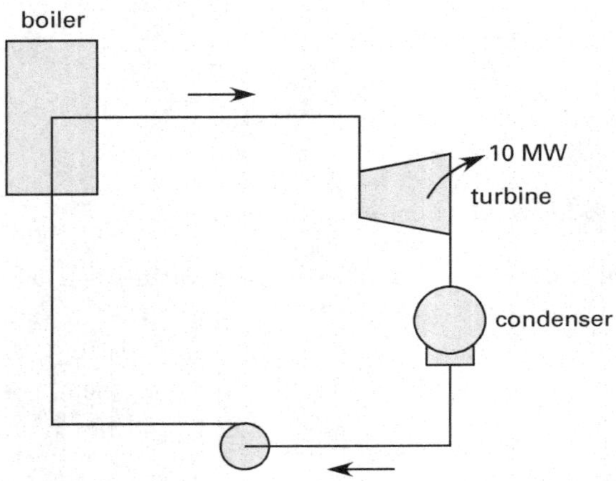

The thermal efficiency of the cycle is most nearly

(A) 9.1%

(B) 28%

(C) 31%

(D) 34%

Hint: Sufficient accuracy can be obtained by neglecting pump work.

PROBLEM 15

A steam turbine generating system that operates at an overall efficiency of 75% with a back pressure of 300 psia is required to generate 9 MW. Superheated steam at 1300 psia and 940°F is used. The amount of steam required is most nearly

(A) 66.7×10^3 lbm/hr

(B) 113×10^3 lbm/hr

(C) 171×10^3 lbm/hr

(D) 228×10^3 lbm/hr

Hint: A turbine can be modeled as an isentropic process.

PROBLEM 16

140 gpm of water at 180°F is pumped out of a vented tank. For this pump at this flow, the manufacturer requires 5.0 ft of net positive suction head (NPSH).

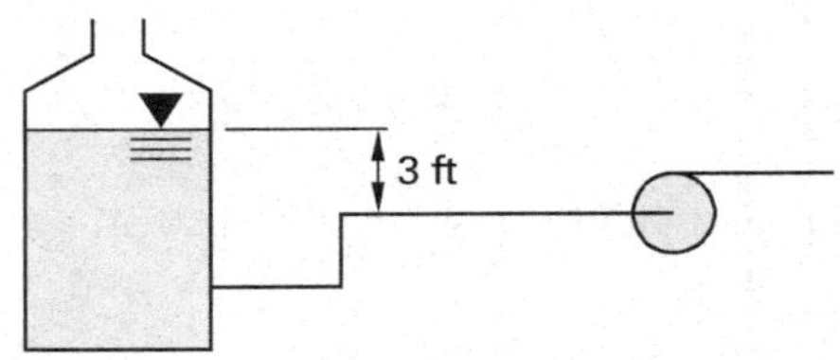

The head loss between the tank and the pump suction is given by the equation

$$h_{f(s)} = 1.50 \times 10^{-5} L Q^{1.852}$$

$h_{f(s)}$ is in feet, L is the pipeline length in feet, and Q is the flow rate in gallons per minute. Surroundings are at 1.0 atm. The maximum permissible pipeline length, if cavitation is to be avoided, is most nearly

(A) 85 ft

(B) 100 ft

(C) 140 ft

(D) 230 ft

Hint: Compare required NPSH and available NPSH.

PROBLEM 17

An impeller with a 2.0 ft diameter operates at its best efficiency point. At this efficiency point, its speed is 440 rpm and it produces a flow rate of 200 gpm, a head of 180 ft, and 17.0 hp of power. If the impeller is clipped to 1.5 ft and still operates at 440 rpm, the new flow rate, head, and power will be most nearly

(A) 150 gpm, 140 ft, and 13 hp

(B) 200 gpm, 140 ft, and 9.6 hp

(C) 84 gpm, 76 ft, and 7.2 hp

(D) 150 gpm, 100 ft, and 7.2 hp

Hint: The impeller is clipped; it is not geometrically similar.

PROBLEM 18

A fan rated at 4700 standard cubic feet per minute (SCFM) is placed in 70°F air at an altitude of 6000 ft. The actual airflow rate of the fan at this temperature and altitude is most nearly

(A) 3800 ft^3/min

(B) 5900 ft^3/min

(C) 6100 ft^3/min

(D) 9200 ft^3/min

Hint: A density correction factor for altitude will be needed.

PROBLEM 19

The flow characteristics of a typical centrifugal compressor are shown.

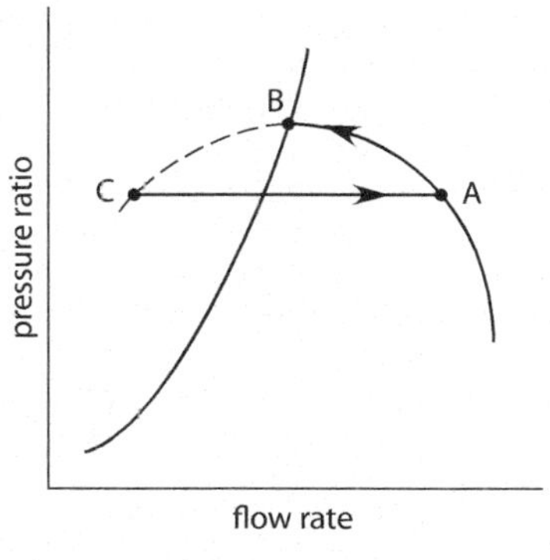

The cycle ABCA is the phenomenon known as

(A) surging

(B) stalling

(C) starving

(D) throttling

Hint: Look carefully at the flow rate during this cycle.

PROBLEM 20

A pump running at 3550 rpm must supply 340 ft of head. To run the pump at 75% efficiency, the flow rate must be most nearly

(A) 200 gpm

(B) 500 gpm

(C) 1000 gpm

(D) greater than is possible given the operating conditions

Hint: The specific speed of the pump will be useful.

PROBLEM 21

A 44 in long, 1.25 in diameter compressor shaft is positioned in vee blocks. A dial indicator is placed on the shaft to determine the shaft runout. The readings of the indicator are given in the following table.

angle (deg)	reading (0.1 mil)	angle (deg)	reading (0.1 mil)
0	0.0	190	1.0
10	0.0	200	0.5
20	0.0	210	0.5
30	0.5	220	0.5
40	0.5	230	0.0
50	0.5	240	0.0
60	0.5	250	0.0
70	1.0	260	–0.5
80	1.0	270	–0.5
90	1.0	280	–0.5
100	1.0	290	–0.5
110	1.0	300	–1.0
120	1.5	310	–1.0
130	1.5	320	–0.5
140	1.5	330	–0.5
150	1.5	340	–0.5
160	1.5	350	0.0
170	1.0	360	0.0
180	1.0		

The mechanical runout of the shaft is most nearly

(A) 0.036 mil

(B) 0.10 mil

(C) 0.15 mil

(D) 0.25 mil

Hint: The mechanical runout is a difference.

PROBLEM 22

A fluid leaving the impeller of a 24 in diameter centrifugal compressor must have a tangential velocity of 300 ft/sec to be properly compressed. The mass flow rate is 900 lbm/sec and the operating speed of the compressor is 6300 rpm. The horsepower of the compressor is most nearly

(A) 170 hp

(B) 1600 hp

(C) 5700 hp

(D) 10,000 hp

Hint: Assume a single-stage compressor to get an estimate of the power.

PROBLEM 23

A three-phase motor operating at 93% efficiency and pulling 100 A is driving a high-pressure boiler feedwater pump in a refinery that pumps 500 gpm with an 850 psi differential. If the motor voltage is 2300 V and the power factor is 0.87, the pump efficiency is most nearly

(A) 57%

(B) 77%

(C) 86%

(D) 99%

Hint: When efficiencies are given, begin looking at power.

PROBLEM 24

A run-around heat recovery system uses a coolant loop and two heat exchangers to extract energy from warm air exhausted from a building and transfer the energy to cold air brought into the building.

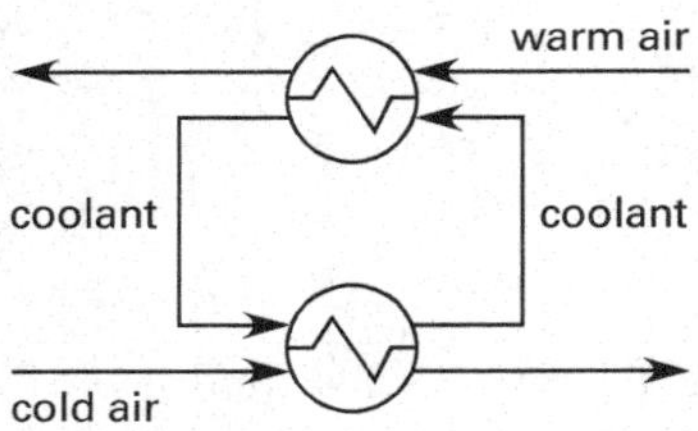

Before passing through the system, the warm air is at 75°F and the cold air is at 10°F. Air flows through the system in both directions at 36,000 lbm/hr, and cooling flows at 15,000 lbm/hr. Heat is removed from the outgoing air at 200,000 Btu/hr.

The overall heat transfer coefficient for each heat exchanger is 50 Btu/hr-ft^2-°F. The heat exchangers are a crossflow type with both fluids unmixed. The specific heats are 0.24 Btu/lbm-°F for the air and 1.0 Btu/lbm-°F for the coolant. Heat loss from the coolant line and energy added by the coolant pump can be neglected.

The coolant temperatures in the left and right halves, respectively, of the loop are most nearly

(A) 75°F and 10°F

(B) 49°F and 36°F

(C) 54°F and 30°F

(D) 30°F and 54°F

Hint: Due to the symmetry of this arrangement, the average of the coolant temperatures on the two sides of the loop is equal to the average of the temperatures of the incoming warm and cold air.

PROBLEM 25

A centrifugal pump is to deliver 300 gpm of hot water from an elevated tank. The water in the tank is saturated liquid at 120 psig and is delivered to the pump through 12 in schedule-40 pipe. The pressure at the pump inlet is 200 psig. The critical cavitation number is 1.2. The flow velocity inside the pump at which cavitation will begin is most nearly

(A) 75 ft/sec

(B) 110 ft/sec

(C) 170 ft/sec

(D) 220 ft/sec

Hint: The vapor pressure will be needed.

PROBLEM 26

A 70,000 lbm steel waste-heat boiler contains 45,000 lbm of water at a temperature of 90°F. The gases at 1300°F enter the unit at a rate of 100,000 lbm/hr. The boiler surface area is 30,000 ft^2. The flue gas specific heat is 0.25 Btu/lbm-°F. The overall heat transfer coefficient is 5 Btu/ft^2-hr-°F. The minimum time required to bring the boiler to 212°F is most nearly

(A) 12 min

(B) 14 min

(C) 23 min

(D) 31 min

Hint: This is similar to a heat exchanger problem.

PROBLEM 27

A cooling tower design requires a water area of 608 ft^2. Since this size is not available, a two-cell tower having a water area of 576 ft^2 (2 times 12 ft by 24 ft) will be used. The increase in the air rate needed to handle the increase in the water load for this two-cell tower is most nearly

(A) 2.1%

(B) 5.2%

(C) 6.0%

(D) 15%

Hint: A correlation equation based on experimental data will be useful.

PROBLEM 28

A noncooled main air blower at a refinery compresses 50,000 cfm of atmospheric air to 53 psia. The compressor efficiency is 79%. The motor shaft horsepower required to operate the main air blower is most nearly

(A) 110 hp

(B) 1800 hp

(C) 4000 hp

(D) 6300 hp

Hint: Think adiabatic power.

PROBLEM 29

A compressor is needed to compress natural gas from 35 psig to 200 psig, at a temperature of 60°F and a rate of 1.7×10^6 standard cubic feet per 24 hours. The compressor will be located 3000 ft above sea level, and the ratio of specific heat of the gas is 1.28. The horsepower of the compressor must be most nearly

(A) 95 hp

(B) 120 hp

(C) 160 hp

(D) 200 hp

Hint: The compression ratio is important.

PROBLEM 30

A centrifugal pump is placed above a large reservoir and will be used to pump 10 gpm of water. To prevent cavitation at this flow rate, the pump manufacturer specifies that the pump inlet pressure should be 20 psi. The water temperature is 70°F, and atmospheric pressure is 14.7 psi. The net positive suction head required is most nearly

(A) 3.8 ft

(B) 20 ft

(C) 46 ft

(D) 80 ft

Hint: Remember to do all the conversions needed.

PROBLEM 31

An AC three-phase generator supplies 20 A at 240 V with a lagging power factor of 0.8. The generator efficiency is 90%. The torque needed to drive the generator at 1800 rpm is most nearly

(A) 17 ft-lbf

(B) 29 ft-lbf

(C) 37 ft-lbf

(D) 50 ft-lbf

Hint: Be sure to realize that this is a three-phase generator.

PROBLEM 32

Desert air at 115°F and 10% relative humidity is to be cooled using an evaporative cooler. The minimum temperature to which the air can be cooled is most nearly

(A) 35°F

(B) 45°F

(C) 71°F

(D) 89°F

Hint: A chart from ASHRAE will help.

PROBLEM 33

An axial flow pump is to deliver 3000 gpm when operating against 100 ft of head. The speed at which the pump should be run is most nearly

(A) 1100 rpm

(B) 1400 rpm

(C) 3900 rpm

(D) 4300 rpm

Hint: Pump efficiency is a key consideration.

PROBLEM 34

A normal shock occurs in air flowing at Mach 2.2. The change in entropy across the shock wave is most nearly

(A) 0.032 Btu/lbm-°R

(B) 0.18 Btu/lbm-°R

(C) 0.27 Btu/lbm-°R

(D) 0.37 Btu/lbm-°R

Hint: Pay attention to upstream and downstream conditions.

PROBLEM 35

Helium at 293K and 101 kPa (absolute) enters a duct at Mach 0.1. The duct is insulated and has a constant cross-sectional area. No external work is being done on the system, but friction is present. The T/T^* ratio is most nearly

(A) 0.56

(B) 1.0

(C) 1.3

(D) 2.6

Hint: The flow is not isentropic.

PROBLEM 36

Three oil derricks each produce the same number of barrels of oil per day. The output from each derrick is sent to a main transportation line by way of a feeder line constructed of 16 in schedule-40 pipe.

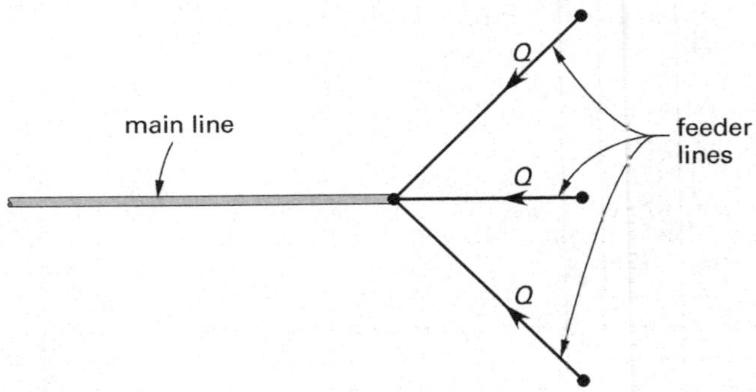

The main line is constructed of 30 in schedule-40 pipe. Elevation effects are negligible, and the friction factors in all the pipes are equal. The ratio of the pressure drop per unit length in the main transportation line to that in the feeder lines is most nearly

(A) 0.04

(B) 0.33

(C) 1.4

(D) 3.0

Hint: Don't forget about the flow rate equation $Q = VA$.

SOLUTION 1

This problem is solved using the pump affinity laws.

$$\frac{Q_1}{Q_2} = \frac{n_1}{n_2}$$

$$\frac{\text{BHP}_1}{\text{BHP}_2} = \left(\frac{n_1}{n_2}\right)^3$$

Substitute the flow ratio for the speed ratio in the brake horsepower equation, substitute the given values, and solve for the new brake horsepower, BHP_2.

$$\frac{\text{BHP}_1}{\text{BHP}_2} = \left(\frac{Q_1}{Q_2}\right)^3$$

$$\frac{150 \text{ hp}}{\text{BHP}_2} = \left(\frac{1700 \text{ gpm}}{2500 \text{ gpm}}\right)^3$$

$$\text{BHP}_2 = 477 \text{ hp} \quad (480 \text{ hp})$$

The answer is (D).

Why Other Options Are Wrong

(A) This answer is incorrect. The brake horsepower of the pump is a function of the pump's flow rate. Therefore, an increase in the flow rate requires an increase in the brake horsepower.

(B) This incorrect answer results when the flow rate ratio is not cubed in the calculation of the brake horsepower.

(C) This incorrect answer results when the flow rate ratio is squared instead of cubed in the calculation of the brake horsepower.

SOLUTION 2

The pump's efficiency is found using a graph that gives the pump efficiency as a function of specific speed and capacity. The specific speed is a dimensionless design parameter used to describe and classify pump impellers. The specific speed is found using the formula

$$n_s = \frac{n\sqrt{Q}}{h^{0.75}}$$

A dimensional analysis of the specific speed equation will show that n_s is not truly dimensionless. Converting Q to cubic feet per minute and multiplying h by the acceleration of gravity, g, before raising it to the 0.75 power will make the equation dimensionless. The convention in the centrifugal pump industry, however, is not to include these constant terms.

The speed of the 60 Hz, two-pole motor needed to drive the pump can be found using the equation

$$n = \frac{7200 \text{ rpm}}{N} = \frac{7200 \text{ rpm}}{2} = 3600 \text{ rpm}$$

The true operating speed of a two-pole motor ranges from 3450 to 3550 rpm. The reduced speed is due to slippage between the rotor and stator.

Using 3550 rpm as the speed, the specific speed for this pump is

$$n_s = \frac{(3550 \text{ rpm})\sqrt{100 \ \frac{\text{gal}}{\text{min}}}}{(240 \text{ ft})^{0.75}} = 582$$

Using a "typical" pump efficiency graph, the pump's maximum attainable efficiency is found to be 52%. Since pump efficiency data is manufacturer-specific, variations in this result can be expected.

The answer is (C).

Why Other Options Are Wrong

(A) This incorrect answer results when the 100 gpm is converted to cubic feet per minute.

(B) This incorrect answer results when the speed of a four-pole motor, 1780 rpm, is used instead the speed of a two-pole motor.

(D) This incorrect answer results when the efficiency is read from the 1000 gpm curve instead of the 100 gpm curve.

SOLUTION 3

The pump's efficiency is the ratio of the pump's water horsepower (WHP) to the pump's brake horsepower (BHP).

$$\eta = \frac{\text{WHP}}{\text{BHP}}$$

The brake horsepower can be found using the pump's shaft torque and speed.

$$\begin{aligned}\text{BHP} &= T\omega \\ &= (1500 \text{ ft-lbf})\left(1750 \ \frac{\text{rev}}{\text{min}}\right)\left(2\pi \ \frac{\text{rad}}{\text{rev}}\right)\left(\frac{1 \text{ min}}{60 \text{ sec}}\right) \\ &\quad \times \left(\frac{1 \text{ hp}}{550 \ \frac{\text{ft-lbf}}{\text{sec}}}\right) \\ &= 500 \text{ hp}\end{aligned}$$

The water horsepower can be found using the formula

$$\text{WHP} = Q\gamma h_A$$

The flow rate is

$$Q = \left(70{,}000\ \frac{\text{barrels}}{\text{day}}\right)\left(42\ \frac{\text{gal}}{\text{barrel}}\right)\left(\frac{1\ \text{ft}^3}{7.48\ \text{gal}}\right) \times \left(\frac{1\ \text{day}}{86{,}400\ \text{sec}}\right)$$
$$= 4.549\ \text{ft}^3/\text{sec}$$

The water horsepower is

$$\begin{aligned}\text{WHP} &= Q\gamma(\text{SG})h_A \\ &= \left(4.549\ \frac{\text{ft}^3}{\text{sec}}\right)\left(62.4\ \frac{\text{lbf}}{\text{ft}^3}\right)(0.86)(956\ \text{ft}) \times \left(\frac{1\ \text{hp}}{550\ \frac{\text{ft-lbf}}{\text{sec}}}\right) \\ &= 424\ \text{hp}\end{aligned}$$

The efficiency of the pump is

$$\eta = \frac{424\ \text{hp}}{500\ \text{hp}} = 0.848 \quad (85\%)$$

The answer is (C).

Why Other Options Are Wrong

(A) This incorrect answer results when the 70,000 barrels per day is converted to gallons per second and then the reciprocal of the efficiency is used.

(B) This incorrect answer results when the specific gravity of crude oil is not included in the WHP calculation. Therefore, the water horsepower is calculated for water instead of crude oil.

(D) This incorrect answer results when 0.68 is used for the specific gravity instead of 0.86.

SOLUTION 4

The friction head at the design point is

$$h_{f,d} = \frac{fL\text{v}_d^2}{2dg} = K\text{v}_d^2 = K\left(\frac{\dot{m}_d}{\rho A}\right)^2$$

The friction head at the off-design point is

$$h_{f,\text{off}} = \frac{fL\text{v}_{\text{off}}^2}{2dg} = K\left(\frac{\dot{m}_{\text{off}}}{\rho A}\right)^2$$

Solving these two equations for the pressure head at the off-design conditions gives

$$h_{f,\text{off}} = h_{f,d}\frac{\dot{m}_{\text{off}}^2}{\dot{m}_d^2}$$

The pressure drop at the off-design is

$$\begin{aligned}\Delta p_{\text{off}} &= \Delta p_d\frac{\dot{m}_{\text{off}}^2}{\dot{m}_d^2} \\ &= \left(33\ \frac{\text{lbf}}{\text{in}^2}\right)\frac{\left(740{,}000\ \frac{\text{lbm}}{\text{hr}}\right)^2}{\left(1.4\times 10^6\ \frac{\text{lbm}}{\text{hr}}\right)^2} \\ &= 9.22\ \text{lbf/in}^2 \quad (9.2\ \text{lbf/in}^2)\end{aligned}$$

The answer is (A).

Why Other Options Are Wrong

(B) This incorrect answer results when the exponent on the mass flow rate is neglected.

(C) This incorrect answer results when the pressure drop for all operating conditions is incorrectly assumed to be equal to the design pressure drop, and therefore independent of the mass flow rate.

(D) This incorrect answer results when the proportionality equation is inverted as

$$\Delta p_{\text{off}} = \Delta p_d\frac{\dot{m}_d^2}{\dot{m}_{\text{off}}^2}$$

SOLUTION 5

A static column of liquid of height h (feet of head) is equivalent to a certain pressure. The relationship between the two is given by

$$p = h\gamma(\text{SG})$$

The pressure at the discharge for a head of 450 ft of crude oil is

$$\begin{aligned}p &= (450\ \text{ft})\left(62.4\ \frac{\text{lbf}}{\text{ft}^3}\right)\left(\frac{1\ \text{ft}}{12\ \text{in}}\right)^2(0.86) \\ &= 168\ \text{lbf/in}^2 \quad (170\ \text{psi})\end{aligned}$$

Note that head and pressure are not interchangeable terms since head is potential energy and pressure is force per unit area.

The answer is (A).

Why Other Options Are Wrong

(B) This incorrect answer results when the specific gravity of crude oil is not included in the calculation of γ. Therefore, this is the pressure if the liquid being pumped were water.

(C) This incorrect answer results when 1 ft^2 is converted to 12 in^2 instead of $(12\text{ in})^2$.

(D) This incorrect answer results when pressure is calculated in pounds per square foot, not pounds per square inch.

SOLUTION 6

The p-V diagram for a Carnot cycle is shown.

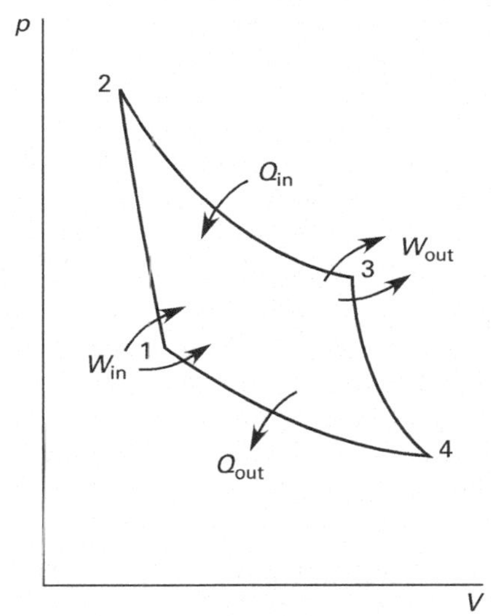

The maximum pressure and temperature occur at state 2, and the minimum pressure and temperature occur at state 4. Therefore, state 2 is where the minimum cylinder volume occurs, and state 4 is where the maximum cylinder volume occurs. The ideal gas law can be used to find the volume at state 2. Then the cycle process can be followed from state 2 to state 3, and then to state 4 where the maximum cylinder volume can be found.

At state 2,

$$p_2V_2 = mRT_2$$

$$\left(1300\ \frac{\text{lbf}}{\text{in}^2}\right)\left(12\ \frac{\text{in}}{\text{ft}}\right)^2 V_2 = (0.2\text{ lbm})\left(53.35\ \frac{\text{ft-lbf}}{\text{lbm-°R}}\right) \times (1200\text{°F} + 460\text{°})$$

$$V_2 = 0.095\text{ ft}^3$$

The heat added to the air is known, so the volume at state 3 can be found using the equation

$$q_{\text{in}} = p_2V_2 \ln \frac{V_3}{V_2}$$

$$(5\text{ Btu})\left(778\ \frac{\text{ft-lbf}}{\text{Btu}}\right) = \left(1300\ \frac{\text{lbf}}{\text{in}^2}\right)\left(12\ \frac{\text{in}}{\text{ft}}\right)^2 \times (0.095\text{ ft}^3) \ln \frac{V_3}{0.095\text{ ft}^3}$$

$$V_3 = 0.12\text{ ft}^3$$

The process from state 3 to state 4 is reversible and adiabatic, so V_4 can be found using the equation

$$\frac{V_4}{V_3} = \left(\frac{T_3}{T_4}\right)^{\frac{1}{k-1}}$$

The ratio of specific heats, k, for air is equal to 1.4. Thus, V_4, the maximum cylinder volume, is

$$\frac{V_4}{0.12\text{ ft}^3} = \left(\frac{1200\text{°F} + 460\text{°}}{80\text{°F} + 460\text{°}}\right)^{\frac{1}{1.4-1}}$$

$$V_4 = 2.0\text{ ft}^3$$

The answer is (C).

Why Other Options Are Wrong

(A) This incorrect answer results when the volume at state 2 is thought to be the maximum cylinder volume.

(B) This incorrect answer results when the conversion to Rankine is forgotten when calculating V_2.

(D) This incorrect answer results when the volume at state 2 is calculated using the universal gas constant instead of the specific gas constant and this volume is thought to be the maximum cylinder volume.

SOLUTION 7

For a steady-state process, the first law of thermodynamics gives

$$Q - W + \dot{m}\left(h_1 + \frac{v_1^2}{2g_c} + gz_1\right) - \dot{m}\left(h_2 + \frac{v_2^2}{2g_c} + gz_2\right) = 0$$

Ignoring potential energy changes and solving for the heat added to the compressor,

$$Q = W - \dot{m}\left(h_1 - h_2 + \frac{v_1^2}{2g_c} - \frac{v_2^2}{2g_c}\right)$$

Hydraulic and Fluid Apps.

The mass flow rate for the incoming air is

$$\dot{m}_1 = \frac{A_1 \mathrm{v}_1}{v_1} = \frac{A_1 \mathrm{v}_1 p_1}{R T_1}$$

$$= \frac{(1.0\ \text{ft}^2)\left(20\ \frac{\text{ft}}{\text{sec}}\right)\left(14.7\ \frac{\text{lbf}}{\text{in}^2}\right)}{\left(53.35\ \frac{\text{ft-lbf}}{\text{lbm-}^\circ\text{R}}\right)(537^\circ\text{R})} \times \left(3600\ \frac{\text{sec}}{\text{hr}}\right)\left(12\ \frac{\text{in}}{\text{ft}}\right)^2$$

$$= 5320\ \text{lbm/hr}$$

From property tables for low-pressure air, the enthalpy of the air at the inlet and outlet of the compressor, respectively, are

$$h_1 = 128.34\ \text{Btu/lbm}$$
$$h_2 = 196.69\ \text{Btu/lbm}$$

Substituting into the equation for the heat added to the compressor,

$$Q = (-150\ \text{hp})\left(2545\ \frac{\text{Btu}}{\text{hp-hr}}\right) - \left(5320\ \frac{\text{lbm}}{\text{hr}}\right) \times \left(\begin{array}{l} 128.34\ \frac{\text{Btu}}{\text{lbm}} - 196.69\ \frac{\text{Btu}}{\text{lbm}} \\ + \frac{\left(20\ \frac{\text{ft}}{\text{sec}}\right)^2 - \left(6\ \frac{\text{ft}}{\text{sec}}\right)^2}{(2)\left(32.2\ \frac{\text{ft-lbm}}{\text{lbf-sec}^2}\right)} \\ \times \left(\frac{1\ \text{Btu}}{778\ \text{ft-lbf}}\right) \end{array}\right)$$

$$= -18{,}167\ \text{Btu/hr} \quad (-18{,}000\ \text{Btu/hr})$$

The answer is (C).

Why Other Options Are Wrong

(A) This incorrect solution results from omitting the conversion factor from ft-lbf to Btu.

(B) This incorrect solution results from omitting g_c in the equation.

(D) This answer results from erroneously assuming that vapor compression is an adiabatic process.

SOLUTION 8

Internal combustion engines are evaluated on the basis of an air-standard Otto cycle. The processes involved are

1–2 isentropic compression
2–3 heat addition at constant volume
3–4 isentropic expansion
4–1 heat rejection at constant volume

The thermal efficiency of an Otto cycle can be calculated either from T_1 and T_2 or from T_4 and T_3.

$$\eta_{\text{th}} = 1 - \frac{T_1}{T_2} = 1 - \frac{T_4}{T_3}$$

The temperatures must be in degrees Rankine.

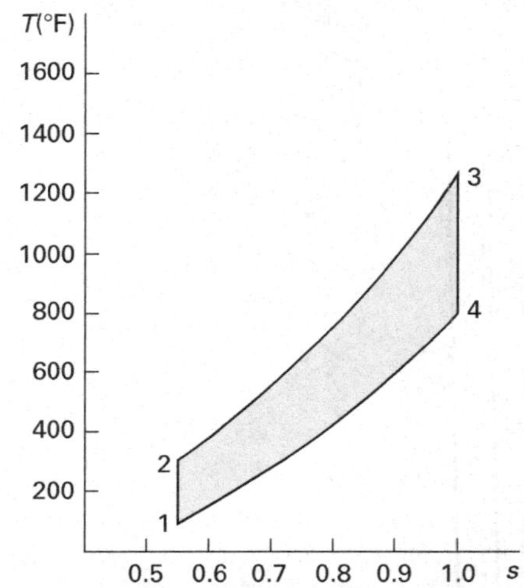

From the T-s diagram, $T_1 = 100°\text{F}$ and $T_2 = 300°\text{F}$. The efficiency is

$$\eta_{\text{th}} = 1 - \frac{T_1}{T_2} = 1 - \frac{100^\circ\text{F} + 460^\circ}{300^\circ\text{F} + 460^\circ} = 1 - \frac{560^\circ\text{R}}{760^\circ\text{R}}$$
$$= 0.26 \quad (26\%)$$

Alternatively,

$$\eta_{\text{th}} = 1 - \frac{T_4}{T_3} = 1 - \frac{800^\circ\text{F} + 460^\circ}{1250^\circ\text{F} + 460^\circ} = 1 - \frac{1260^\circ\text{R}}{1710^\circ\text{R}}$$
$$= 0.26 \quad (26\%)$$

The answer is (A).

Why Other Options Are Wrong

(B) This incorrect answer results when T_3 and T_4 are used but not converted to Rankine.

(C) This incorrect answer results when T_1 and T_2 are used but not converted to Rankine.

(D) This incorrect answer results when the Carnot efficiency is calculated and the temperatures are not converted to Rankine.

SOLUTION 9

The thermal efficiency of a reversible heat engine operating between two reservoirs and exchanging only heat with these reservoirs is given by

$$\eta_{th} = 1 - \frac{T_L}{T_H}$$

T_L and T_H are the low and high temperatures, respectively. From this it can be seen that both statements I and III are true.

The above thermal efficiency represents the theoretical best efficiency that can be achieved. It can be shown that the Carnot cycle achieves this efficiency, so statement II is also true.

Statement IV is false because it can be shown that different reversible engines operating between the same temperature limits will have the same thermal efficiencies.

The answer is (D).

Why Other Options Are Wrong

(A) This is incorrect because statement III is also true.

(B) This is incorrect because statements II and III are also true and statement IV is false.

(C) This is incorrect because statement I is also true.

SOLUTION 10

First determine the brake horsepower needed for the original pump design. Using the performance curves, the brake horsepower for 350 gpm and 200 ft of head is approximately 28 bhp.

Now determine the brake horsepower for the oversized pump. The oversized pump rating is

$$\begin{aligned} Q_{os} &= (\text{oversize factor})Q = (1.2)\left(350\ \frac{\text{gal}}{\text{min}}\right) \\ &= 420\ \text{gpm} \\ h_{os} &= (\text{oversize factor})h = (1.2)(200\ \text{ft}) \\ &= 240\ \text{ft} \end{aligned}$$

Using the performance curves, the brake horsepower for 420 gpm and 240 ft of head is approximately 40 bhp.

The larger pump will have to be throttled back to a flow rate of 350 gpm once put into service. Both pumps deliver 350 gpm, but the oversized pump needs an additional 12 hp to do it. This 12 hp is wasted energy, and the additional yearly electrical cost is

$$\begin{aligned} &(12\ \text{hp})\left(0.746\ \frac{\text{kW}}{\text{hp}}\right)\left(\frac{\$0.07}{\text{kW-hr}}\right)\left(24\ \frac{\text{hr}}{\text{day}}\right)\left(365\ \frac{\text{days}}{\text{yr}}\right) \\ &\quad = \$5489/\text{yr} \quad (\$5490/\text{yr}) \end{aligned}$$

The answer is (D).

Why Other Options Are Wrong

(A) For this pump, a larger impeller diameter is needed to handle the 20% additional load. Thus, more power is needed, so the pump as originally designed cannot handle the additional load.

(B) This incorrect answer results when the brake horsepower of the original design is misread as 38 bhp.

(C) This incorrect answer results when, in oversizing the pump, only the flow rate is increased and the head increase is forgotten.

SOLUTION 11

This problem is solved using the pump affinity laws.

$$\frac{Q_1}{Q_2} = \frac{n_1}{n_2}$$

$$\frac{h_1}{h_2} = \left(\frac{n_1}{n_2}\right)^2$$

With the given curves, the only way to solve this problem is through trial and error. The known values are

$$\begin{aligned} Q_2 &= 550\ \text{gpm} \\ h_2 &= 350\ \text{ft} \\ n_1 &= 3550\ \text{rpm} \end{aligned}$$

Since n_2 is not known, a speed must be guessed. If the resulting values for Q_1 and h_1 lie on the 9.5 in impeller curve, then the speed is correct. If not, a new guess must be tried.

Guess 1: n_2 is 3800 rpm.

$$Q_1 = \left(\frac{3550\ \frac{\text{rev}}{\text{min}}}{3800\ \frac{\text{rev}}{\text{min}}}\right)\left(550\ \frac{\text{gal}}{\text{min}}\right) = 514\ \text{gpm}$$

$$h_1 = \left(\frac{3550\ \frac{\text{rev}}{\text{min}}}{3800\ \frac{\text{rev}}{\text{min}}}\right)^2 (350\ \text{ft}) = 305\ \text{ft}$$

This point lies near the $10\frac{3}{16}$ in diameter impeller curve, well above the 9.5 in diameter impeller curve.

Guess 2: n_2 is 4100 rpm.

$$Q_1 = \left(\frac{3550\ \frac{\text{rev}}{\text{min}}}{4100\ \frac{\text{rev}}{\text{min}}}\right)\left(550\ \frac{\text{gal}}{\text{min}}\right) = 476\ \text{gpm}$$

$$h_1 = \left(\frac{3550\ \frac{\text{rev}}{\text{min}}}{4100\ \frac{\text{rev}}{\text{min}}}\right)^2 (350\ \text{ft}) = 262\ \text{ft}$$

This point lies close to the 9.5 in impeller curve, so the speed to produce 550 gpm and 350 ft of head is approximately 4100 rpm.

Guess 3: n_2 is 4150 rpm.

$$Q_1 = \left(\frac{3550\ \frac{\text{rev}}{\text{min}}}{4150\ \frac{\text{rev}}{\text{min}}}\right)\left(550\ \frac{\text{gal}}{\text{min}}\right) = 470\ \text{gpm}$$

$$h_1 = \left(\frac{3550\ \frac{\text{rev}}{\text{min}}}{4150\ \frac{\text{rev}}{\text{min}}}\right)^2 (350\ \text{ft}) = 256\ \text{ft}$$

This point lies right on the 9.5 in impeller curve. So the required speed is 4150 rpm.

The answer is (B).

Why Other Options Are Wrong

(A) This incorrect answer results when, in the calculation of h_1, the n_1 and n_2 speeds in the speed ratio are reversed.

(C) This incorrect answer results when, in the calculation of h_1, the speed ratio is not squared.

(D) This incorrect answer results when, in the calculation of h_1, 550 gpm is used instead of the 350 ft head.

SOLUTION 12

This problem can be solved using the power and torsional stress equations.

$$P = T\omega$$

$$\tau = \frac{Tr}{J}$$

Using the maximum shear stress failure criterion and a yield stress of 30,000 psi for the steel shaft the maximum allowable shear stress in the shaft is

$$\tau = 0.5\sigma_y = (0.5)\left(30{,}000\ \frac{\text{lbf}}{\text{in}^2}\right) = 15{,}000\ \text{psi}$$

The maximum torque is found from the equation

$$\tau = \frac{Tr}{J}$$

$$15{,}000\ \frac{\text{lbf}}{\text{in}^2} = \frac{T(1.25\ \text{in})}{\left(\frac{\pi}{32}\right)d^4} = \frac{T(1.25\ \text{in})}{\left(\frac{\pi}{32}\right)(2.5\ \text{in})^4}$$

$$T = 46{,}000\ \text{lbf-in}$$

The maximum power that can be transmitted is

$$\begin{aligned} P &= T\omega \\ &= (46{,}000\ \text{lbf-in})\left(\frac{1\ \text{ft}}{12\ \text{in}}\right)\left(200\ \frac{\text{rev}}{\text{min}}\right)\left(2\pi\ \frac{\text{rad}}{\text{rev}}\right) \\ &\quad \times \left(\frac{1\ \text{min}}{60\ \text{sec}}\right)\left(\frac{1\ \text{hp}}{550\ \frac{\text{lbf-ft}}{\text{sec}}}\right) \\ &= 146\ \text{hp} \quad (150\ \text{hp}) \end{aligned}$$

The answer is (B).

Why Other Options Are Wrong

(A) This incorrect answer results from not converting units of pound-feet per second to horsepower and adjusting the decimal point to bring the answer into a reasonable range.

(C) This incorrect answer results when the cross-sectional area is used instead of the polar area moment of inertia.

(D) This incorrect answer results when 30,000 psi is used as the maximum shear stress instead of 15,000 psi.

SOLUTION 13

The operating point of a pump is where the system curve crosses the pump curve. The system curve can be found, using the affinity laws, from the following equation and by developing a table of values for Q_1 versus h_1.

$$\frac{h_1}{h_2} = \left(\frac{Q_1}{Q_2}\right)^2$$

$$h_{1,\text{ft}} = h_{2,\text{ft}}\left(\frac{Q_{1,\text{gpm}}}{Q_{2,\text{gpm}}}\right)^2$$

$$= 200\left(\frac{Q_{1,\text{gpm}}}{500}\right)^2$$

$$= (8 \times 10^{-4})Q_{1,\text{gpm}}^2$$

Q_1 (gpm)	h_1 (ft)
0	0
200	32
400	128
600	288
800	512
1000	800

Plotting these points on the given graph, it can be seen that the system curve crosses the pump curve at approximately $Q = 800$ gpm. (A more exact answer is $Q = 785$ gpm.)

The answer is (C).

Why Other Options Are Wrong

(A) This incorrect answer results from switching 200 and 500 in the equation for $h_{1,\text{ft}}$.

(B) This incorrect answer results from using the point where the pump efficiency curve crosses the pump curve.

(D) This incorrect answer results from neglecting to square the ratio Q_1/Q_2.

SOLUTION 14

The cycle thermal efficiency is

$$\eta_{\text{th}} = \frac{W_{\text{turbine}} - W_{\text{pump}}}{Q_{\text{in}}}$$

Neglecting pump work, this simplifies to

$$\eta_{\text{th}} = \frac{W_{\text{turbine}}}{Q_{\text{in}}}$$

The work of the turbine can be found from the turbine output and the mass flow rate.

$$W_{\text{turbine}} = \frac{W_{\text{out}}}{\dot{m}}$$

$$= \frac{(10\text{ MW})\left(10^6\ \frac{\text{W}}{\text{MW}}\right)\left(3.413\ \frac{\text{Btu}}{\text{hr-W}}\right)}{10^5\ \frac{\text{lbm}}{\text{hr}}}$$

$$= 341.3\text{ Btu/lbm}$$

The heat transferred into the system, again neglecting the work of the pump, equals the enthalpy of the saturated steam leaving the boiler minus the enthalpy of the saturated liquid leaving the condenser.

$$Q_{\text{in}} = h_{\text{steam}} - h_{\text{liquid}}$$

In the condenser, the condensate is subcooled 6°F. The saturation temperature for water at 2 psia is 126.04°F, so the temperature after subcooling is 120.04°F. From steam tables, the enthalpy of this saturated liquid is 88.00 Btu/lbm. Also from steam tables, the enthalpy of the saturated steam leaving the boiler at 1000 psia is 1192.4 Btu/lbm. Again neglecting pump work,

$$\eta_{\text{th}} = \frac{W_{\text{turbine}}}{Q_{\text{in}}} = \frac{W_{\text{turbine}}}{h_{\text{steam}} - h_{\text{liquid}}}$$

$$= \frac{341.3\ \frac{\text{Btu}}{\text{lbm}}}{1192.4\ \frac{\text{Btu}}{\text{lbm}} - 88.0\ \frac{\text{Btu}}{\text{lbm}}}$$

$$= 0.31 \quad (31\%)$$

The answer is (C).

Why Other Options Are Wrong

(A) This incorrect solution results from leaving out the conversion from watts to Btu/hr.

(B) This incorrect solution results from adjusting the turbine work by multiplying by the turbine efficiency.

(D) This incorrect solution results from adjusting the turbine work by dividing by the turbine efficiency.

SOLUTION 15

The power generated by the turbine can be found using

$$P = \eta_t \dot{m}\Delta h = \eta_t \dot{m}(h_{\text{inlet}} - h_{\text{exhaust}})$$

The turbine efficiency is η_t. Assuming the expansion is isentropic, the entropy at the turbine inlet and exhaust will be the same. This will be used to determine the enthalpy at the exhaust. Using steam tables at 1300 psia and 940°F,

$$h_{\text{inlet}} = 1460.8 \text{ Btu/lbm}$$
$$s_{\text{inlet}} = 1.5941 \text{ Btu/lbm-°F}$$

Assuming isentropic expansion, s_{exhaust} is equal to 1.5941 Btu/lbm-°F. Using this value and a pressure of 300 psia, the steam tables give an enthalpy of

$$h_{\text{exhaust}} = 1281 \text{ Btu/lbm}$$

No interpolation is needed since the entropy is 1.5940 Btu/lbm-°F, which is close enough to s_{inlet}.

Substituting into the power equation,

$$9 \times 10^6 \text{ W} = 0.75\dot{m}\left(1460.8 \ \frac{\text{Btu}}{\text{lbm}} - 1281 \ \frac{\text{Btu}}{\text{lbm}}\right) \times \left(\frac{1 \text{ W}}{3.413 \ \frac{\text{Btu}}{\text{hr}}}\right)$$
$$\dot{m} = 228 \times 10^3 \text{ lbm/hr}$$

The answer is (D).

Why Other Options Are Wrong

(A) This incorrect answer results when the conversion from watts to Btu/hr is neglected.

(B) This incorrect answer results when the internal energy of the superheated steam is used instead of the enthalpy at 1300 psia and 940°F.

(C) This incorrect answer results when the efficiency of the system is ignored.

SOLUTION 16

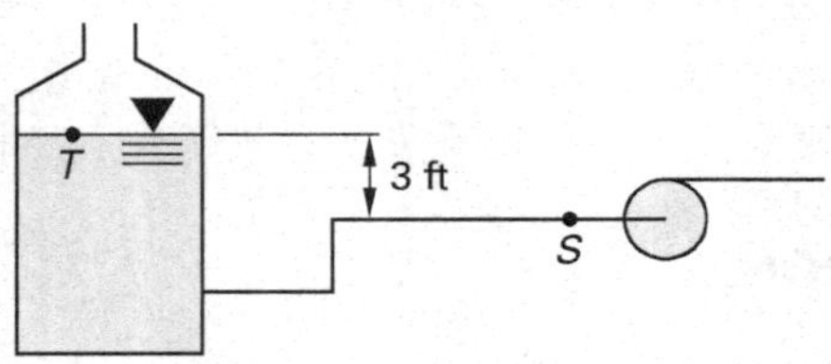

To prevent cavitation, the required NPSH must be less than or equal to the available NPSH, where

$$\text{NPSHA} = (z_T - z_S) + \frac{v_T^2}{2g} + \frac{p_T - p_{\text{sat}}}{\gamma} - h_{f(s)}$$

Solving for $h_{f(s)}$ gives

$$h_{f(s)} < (z_T - z_S) + \frac{v_T^2}{2g} + \frac{p_T - p_{\text{sat}}}{\gamma} - \text{NPSHR}$$

The velocity in the tank is 0 ft/sec, the water level in the tank is 3 ft higher than at the pump intake, atmospheric pressure is 14.7 psia, the saturation pressure of water at 180°F (from steam tables) is 7.52 psia, and the specific weight of water at 180°F is 60.6 lbm/ft^3.

$$h_{f(s)} < (3 \text{ ft} - 0 \text{ ft}) + 0 \text{ ft} + \frac{\left(14.7 \ \frac{\text{lbf}}{\text{in}^2} - 7.52 \ \frac{\text{lbf}}{\text{in}^2}\right)\left(12 \ \frac{\text{in}}{\text{ft}}\right)^2}{60.6 \ \frac{\text{lbf}}{\text{ft}^3}} - 5.0 \text{ ft}$$
$$= 15.06 \text{ ft}$$

Solving for L in the manufacturer's equation,

$$h_{f(s)} = 1.50 \times 10^{-5} L Q^{1.852}$$
$$L < \frac{h_{f(s)}}{(1.50 \times 10^{-5}) Q^{1.852}} = \frac{15.06 \text{ ft}}{(1.50 \times 10^{-5})\left(140 \ \frac{\text{gal}}{\text{min}}\right)^{1.852}}$$
$$= 106 \text{ ft} \quad (100 \text{ ft})$$

The answer is (B).

Why Other Options Are Wrong

(A) This incorrect answer results when $z_T - z_S$ is left out of the calculations.

(C) This incorrect answer results when the subtraction of NPSHR is neglected.

(D) This incorrect answer results when p_{sat} is left out of the calculations.

SOLUTION 17

When two impellers of different size operate at the same speed and efficiency, the affinity laws state that

$$\frac{Q_2}{Q_1} = \frac{d_2}{d_1}$$
$$\frac{h_2}{h_1} = \left(\frac{d_2}{d_1}\right)^2$$
$$\frac{P_2}{P_1} = \left(\frac{d_2}{d_1}\right)^3$$

(These laws are only approximate in the case of a clipped impeller, but they are conventionally treated as exact because the difference is usually trivial.)

Thus,

$$\begin{aligned} Q_2 &= Q_1\left(\frac{d_2}{d_1}\right) \\ &= \left(200\ \frac{\text{gal}}{\text{min}}\right)\left(\frac{1.5\ \text{ft}}{2.0\ \text{ft}}\right) \\ &= 150\ \text{gpm} \end{aligned}$$

$$\begin{aligned} h_2 &= h_1\left(\frac{d_2}{d_1}\right)^2 \\ &= (180\ \text{ft})\left(\frac{1.5\ \text{ft}}{2.0\ \text{ft}}\right)^2 = 101\ \text{ft} \quad (100\ \text{ft}) \end{aligned}$$

$$\begin{aligned} P_2 &= P_1\left(\frac{d_2}{d_1}\right)^3 \\ &= (17.0\ \text{hp})\left(\frac{1.5\ \text{ft}}{2.0\ \text{ft}}\right)^3 = 7.2\ \text{hp} \end{aligned}$$

The answer is (D).

Why Other Options Are Wrong

(A) This incorrect answer results when all three ratios are incorrectly thought to equal d_2/d_1.

(B) This answer results when the affinity laws are incorrectly stated as

$$\frac{Q_2}{Q_1} = 1$$

$$\frac{h_2}{h_1} = \frac{d_2}{d_1}$$

$$\frac{P_2}{P_1} = \left(\frac{d_2}{d_1}\right)^2$$

(C) This answer results when all three ratios are incorrectly thought to equal $(d_2/d_1)^3$.

SOLUTION 18

The actual cubic feet per minute, ACFM, is found using the flow rate at standard conditions known as standard cubic feet per minute, SCFM, and the following equation.

$$\text{ACFM} = K_d(\text{SCFM})$$

The constant K_d is a density correction factor for altitude and is given by

$$K_d = \frac{\rho_{\text{std}}}{\rho_{\text{actual}}} = \left(\frac{p_{\text{std}}}{p_{\text{actual}}}\right)\left(\frac{T_{\text{actual}}}{T_{\text{std}}}\right)$$

Most tables give the ratio $p_{\text{actual}}/p_{\text{std}}$. At 6000 ft this ratio is

$$\frac{p_{\text{actual}}}{p_{\text{std}}} = 0.801$$

Convering the air temperature to degrees Rankine, the density correction factor is therefore

$$K_d = (0.801)^{-1}\left(\frac{70°\text{F} + 460°}{70°\text{F} + 460°}\right) = \frac{1}{0.801}$$

The actual flow rate of the fan is

$$\begin{aligned} \text{ACFM} = K_d(\text{SCFM}) &= \left(\frac{1}{0.801}\right)\left(4700\ \frac{\text{ft}^3}{\text{min}}\right) \\ &= 5870\ \text{ft}^3/\text{min} \quad (5900\ \text{ft}^3/\text{min}) \end{aligned}$$

The answer is (B).

Why Other Options Are Wrong

(A) This incorrect answer results from using the ratio of $\rho_{\text{actual}}/\rho_{\text{std}}$ instead of its reciprocal.

(C) This incorrect answer results from reading an incorrect K_d value of 0.772 from the table.

(D) This incorrect answer results from transposing 4 and 7 in the flow rate.

SOLUTION 19

As the flow rate decreases from point A, the compressor performance moves up the pressure-ratio/flow-rate curve to point B. Further reduction in the flow rate causes the compressor to operate at point C, reducing the compressor's pressure capability. As the compressor's performance moves to point C, refrigerant (in the case of refrigeration operation) continues to boil off because of the evaporator's heat load. This builds up the evaporator pressure and decreases the pressure ratio, causing the compressor to shift briefly back to point A where the cycle repeats. This cycle is known as surging and typically occurs when the compressor is operating at less than 35% of the rated capacity.

The answer is (A).

Why Other Options Are Wrong

(B) Stalling is the breakdown of the airflow through a few stages of a compressor.

(C) Starving is caused by lack of air at the inlet of the compressor.

(D) Throttling is the regulation of flow and head.

SOLUTION 20

The flow rate needed to achieve a pump efficiency of 75% can be found using an efficiency versus specific speed graph, such as is found in the *Mechanical Engineering Reference Manual*. In this graph the 75% efficiency line intersects both the 200 gal/min and 500 gal/min curves. To achieve 75% efficiency at 200 gal/min requires a specific speed of 2000, and at 500 gal/min a specific speed of 1000 is required.

The specific speed is given by the formula

$$n_s = \frac{n\sqrt{Q}}{h_A^{0.75}}$$

For a flow rate of 200 gal/min, the specific speed is

$$n_s = \frac{\left(3550\ \frac{\text{rev}}{\text{min}}\right)\sqrt{200\ \frac{\text{gal}}{\text{min}}}}{(340\ \text{ft})^{0.75}} = 634$$

Specific speed is not dimensionless, but the units are meaningless, so specific speed is expressed as a unitless number.

At this flow rate, the specific speed is 634, not the required 2000. Therefore, this is not the correct flow rate. In fact, at this specific speed and flow rate the pump efficiency is approximately 60%.

For a flow rate of 500 gal/min, the specific speed is

$$n_s = \frac{\left(3550\ \frac{\text{rev}}{\text{min}}\right)\sqrt{500\ \frac{\text{gal}}{\text{min}}}}{(340\ \text{ft})^{0.75}} = 1000$$

This flow rate provides the needed specific speed to achieve a pump efficiency of 75%.

The answer is (B).

Why Other Options Are Wrong

(A) This incorrect flow rate only provides a pump efficiency of 60% as shown.

(C) This flow rate was derived using an incorrect total head of 430 ft.

(D) This answer is incorrect because a pump efficiency of 75% can be achieved.

SOLUTION 21

The mechanical runout is the largest difference between any two readings. The smallest reading is −0.1 mil and the largest reading is 0.15 mil. The difference is 0.15 mil − (−0.1 mil) = 0.25 mil.

The answer is (D).

Why Other Options Are Wrong

(A) This incorrect answer is the average of the dial indicator readings, not the runout of the shaft.

(B) This incorrect answer is the largest negative value from zero, not the runout of the shaft.

(C) This incorrect answer is the largest positive value from zero, not the runout of the shaft.

SOLUTION 22

The power can be estimated using

$$P = \left(\frac{\dot{m}\omega}{g_c}\right)(\text{v}_{t,\text{out}} r_o - \text{v}_{t,\text{in}} r_i)$$

The mass flow rate, $\dot{m}$, angular velocity, ω, tangential velocity of the fluid leaving the impeller, $\text{v}_{t,\text{out}}$, and the outer radius of the impeller, r_o, are given. The tangential velocity of the fluid entering the impeller, $\text{v}_{t,\text{in}}$, is usually negligible, so the power equation becomes

$$P = \left(\frac{\dot{m}\omega}{g_c}\right)\text{v}_{t,\text{out}} r_o$$

Substituting given values,

$$P = \left(\frac{\left(900\ \frac{\text{lbm}}{\text{sec}}\right)\left(6300\ \frac{\text{rev}}{\text{min}}\right)\left(\frac{1\ \text{min}}{60\ \text{sec}}\right)\left(2\pi\ \frac{\text{rad}}{\text{rev}}\right)}{32.2\ \frac{\text{ft-lbm}}{\text{lbf-sec}^2}}\right)$$

$$\times\left(300\ \frac{\text{ft}}{\text{sec}}\right)(12\ \text{in})\left(\frac{1\ \text{ft}}{12\ \text{in}}\right)\left(\frac{1\ \text{hp}}{550\ \frac{\text{ft-lbf}}{\text{sec}}}\right)$$

$$= 10{,}000\ \text{hp}$$

The answer is (D).

Why Other Options Are Wrong

(A) This incorrect answer results from dividing by 3600 instead of 60 when converting to rad/sec.

(B) This incorrect answer results from neglecting the factor of 2π in the conversion from rpm to rad/sec.

(C) This incorrect answer results from transposing the 6 and the 3 in the rpm value.

SOLUTION 23

The pump efficiency can be estimated by comparing the motor's power output to the pump's power consumption.

Starting with the motor efficiency, the power output by the motor is

$$\eta_m = \frac{P_{m,\text{out}}}{P_{m,\text{in}}}$$

$$P_{m,\text{out}} = \eta_m P_{m,\text{in}} = \eta_m \sqrt{3}\, VI \cos\phi$$

The motor efficiency and power factor are η_m and $\cos\phi$, respectively.

$$\begin{aligned} P_{m,\text{out}} &= (0.93)\sqrt{3}\,(2300 \text{ V})(100 \text{ A})(0.87)\left(\frac{1 \text{ kW}}{1000 \text{ W}}\right) \\ &= 322 \text{ kW} \end{aligned}$$

Using the pump efficiency, the power needed by the pump, $P_{p,\text{in}}$, in terms of the differential pressure is

$$\begin{aligned} \eta_p &= \frac{P_{p,\text{out}}}{P_{p,\text{in}}} \\ \eta_p P_{p,\text{in}} &= P_{p,\text{out}} = Q\Delta p \\ &= \left(500 \ \frac{\text{gal}}{\text{min}}\right)\left(\frac{1 \text{ ft}^3}{7.48 \text{ gal}}\right)\left(\frac{1 \text{ min}}{60 \text{ sec}}\right)\left(850 \ \frac{\text{lbf}}{\text{in}^2}\right) \\ &\quad \times \left(12 \ \frac{\text{in}}{\text{ft}}\right)^2 \left(1.356 \times 10^{-3} \ \frac{\text{kW}}{\frac{\text{ft-lbf}}{\text{sec}}}\right) \\ P_{p,\text{in}} &= \frac{185 \text{ kW}}{\eta_p} \end{aligned}$$

By equating the output power from the motor with the input power needed by the pump, an approximate pump efficiency can be obtained.

$$322 \text{ kW} = \frac{185 \text{ kW}}{\eta_p}$$

$$\eta_p = 0.57 \quad (57\%)$$

The answer is (A).

Why Other Options Are Wrong

(B) This incorrect answer results from neglecting to make the conversion from horsepower to kilowatts.

(C) This incorrect answer results from neglecting to use the power factor and the square root of three.

(D) This incorrect answer results from neglecting to use the square root of three.

SOLUTION 24

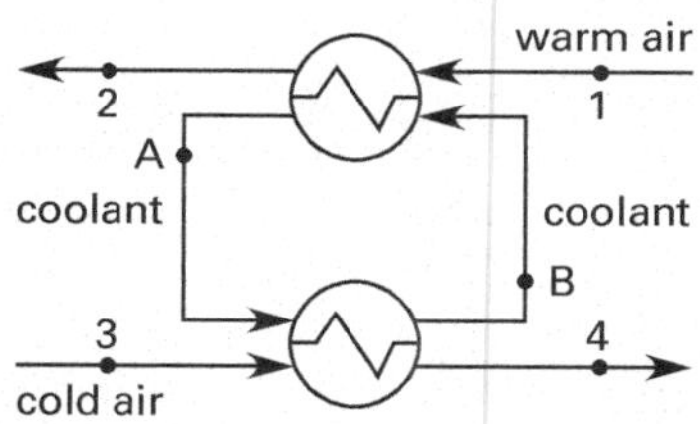

Because of symmetry, the changes in air temperature are equal and opposite.

$$\begin{aligned} T_2 - T_1 &= T_3 - T_4 \\ &= \frac{Q}{(\dot{m}c_p)_{\text{air}}} \\ &= \frac{200{,}000 \ \frac{\text{Btu}}{\text{hr}}}{\left(36{,}000 \ \frac{\text{lbm}}{\text{hr}}\right)\left(0.24 \ \frac{\text{Btu}}{\text{lbm-°F}}\right)} \\ &= 23.2\text{°F} \end{aligned}$$

The changes in coolant temperature are similarly equal and opposite. For the same heat transfer rate,

$$\begin{aligned} T_\text{A} - T_\text{B} &= \frac{Q}{(\dot{m}c_p)_{\text{coolant}}} \\ &= \frac{200{,}000 \ \frac{\text{Btu}}{\text{hr}}}{\left(15{,}000 \ \frac{\text{lbm}}{\text{hr}}\right)\left(1 \ \frac{\text{Btu}}{\text{lbm-°F}}\right)} \\ &= 13.3\text{°F} \end{aligned}$$

The average coolant temperature is

$$\begin{aligned} \overline{T}_{\text{coolant}} &= \frac{T_\text{A} + T_\text{B}}{2} = \frac{T_1 + T_3}{2} \\ &= \frac{10\text{°F} + 75\text{°F}}{2} \\ &= 42.5\text{°F} \end{aligned}$$

To find T_A and T_B,

$$T_A = \overline{T}_{coolant} + \frac{T_A - T_B}{2} = 42.5^\circ F + \frac{13.3^\circ F}{2} = 49.2^\circ F \quad (49^\circ F)$$

$$T_B = \overline{T}_{coolant} - \frac{T_A - T_B}{2} = 42.5^\circ F - \frac{13.3^\circ F}{2} = 35.9^\circ F \quad (36^\circ F)$$

The answer is (B).

Why Other Options Are Wrong

(A) This incorrect answer results when it is assumed that $T_B = T_1$ and $T_A = T_3$.

(C) This incorrect answer results when $(T_2 - T_1)$ is used instead of instead of $(T_A - T_B)$.

(D) This incorrect answer results when $(T_2 - T_1)$ is used instead of $(T_A - T_B)$ and T_A and T_B are interchanged.

SOLUTION 25

The cavitation number, σ, is calculated using the equation

$$\sigma = \frac{2g_c(p - p_{vp})}{\rho v^2}$$

The pressure, p, and the vapor pressure, p_{vp}, are absolute pressures. For no cavitation to occur, the cavitation number, σ, must be greater than the critical cavitation number, σ_{cr}. The vapor pressure is the saturation pressure, so

$$\begin{aligned} p_{vp} &= \text{water pressure} + \text{atmospheric pressure} \\ &= 120\ \frac{\text{lbf}}{\text{in}^2} + 14.7\ \frac{\text{lbf}}{\text{in}^2} \\ &= 134.7\ \text{psia} \end{aligned}$$

To find the correct water density, the temperature of the water must be known. From the saturated steam tables, the temperature of saturated water at 134.7 psia is approximately 350°F. The water density at this temperature is 55.6 lbm/ft³.

Using the critical cavitation number and the equation for σ, the maximum flow velocity with no cavitation can be found.

$$1.2 = \frac{(2)\left(32.2\ \frac{\text{ft-lbm}}{\text{lbf-sec}^2}\right) \times \left(200\ \frac{\text{lbf}}{\text{in}^2} + 14.7\ \frac{\text{lbf}}{\text{in}^2} - 134.7\ \frac{\text{lbf}}{\text{in}^2}\right) \times \left(12\ \frac{\text{in}}{\text{ft}}\right)^2}{\left(55.6\ \frac{\text{lbm}}{\text{ft}^3}\right) v^2}$$

$$v = 105\ \text{ft/sec} \quad (110\ \text{ft/sec})$$

The answer is (B).

Why Other Options Are Wrong

(A) This incorrect answer results when the 2 in the equation for σ is forgotten.

(C) This incorrect answer results when the vapor pressure is not subtracted from the pressure inside the pump.

(D) This incorrect answer results when the equation for σ is calculated with the sum of the two pressures instead of their difference.

SOLUTION 26

Method 1

The following equation can be used to approximate the time needed to increase the temperature of a boiler from T_1 to T_2.

$$\ln\left(\frac{T_{g,\text{in}} - T_1}{T_{g,\text{in}} - T_2}\right) = \left(\frac{\dot{m}_g c_{p,g}\left(e^{\frac{UA}{\dot{m}_g c_p}} - 1\right)}{Ce^{\frac{UA}{\dot{m}_g c_p}}}\right)\Delta t$$

The entering flue gas temperature, $T_{g,\text{in}}$, is 1300°F; the mass flow rate of the flue gases, $\dot{m}_g$, is 100,000 lbm/hr; the flue gas specific heat, $c_{p,g}$, is 0.25 Btu/lbm-°F; the overall heat transfer coefficient, U, is 5 Btu/ft²-hr-°F; and the surface area of the boiler, A, is 30,000 ft².

The thermal capacitance of the boiler, C, must be calculated as follows.

$$\begin{aligned} C &= m_{\text{steel}}c_{p,\text{steel}} + m_{\text{water}}c_{p,\text{water}} \\ &= (70{,}000\ \text{lbm})\left(0.11\ \frac{\text{Btu}}{\text{lbm-°F}}\right) \\ &\quad +(45{,}000\ \text{lbm})\left(1.0\ \frac{\text{Btu}}{\text{lbm-°F}}\right) \\ &= 52{,}700\ \text{Btu/°F} \end{aligned}$$

The exponent of the two exponential terms is

$$\frac{UA}{\dot{m}_g c_p} = \frac{\left(5\ \frac{\text{Btu}}{\text{ft}^2\text{-hr-°F}}\right)(30{,}000\ \text{ft}^2)}{\left(100{,}000\ \frac{\text{lbm}}{\text{hr}}\right)\left(0.25\ \frac{\text{Btu}}{\text{lbm-°F}}\right)} = 6$$

Using these equations and the given values, the time needed to heat the boiler can be calculated.

$$\ln\left(\frac{1300°\text{F} - 90°\text{F}}{1300°\text{F} - 212°\text{F}}\right) = \left(\frac{\left(100{,}000\ \frac{\text{lbm}}{\text{hr}}\right)\left(\frac{1\ \text{hr}}{60\ \text{min}}\right) \times\left(0.25\ \frac{\text{Btu}}{\text{lbm-°F}}\right) \times(e^6 - 1)}{\left(52{,}700\ \frac{\text{Btu}}{°\text{F}}\right)e^6}\right)\Delta t$$

$$\Delta t = 13.5\ \text{min} \quad (14\ \text{min})$$

The answer is (B).

Method 2

The solution can also be obtained through the fundamental concepts used to derive the equation used in Method 1. First, the energy balance of the system can be stated this way: The change in internal energy of the boiler during dt is equal to the net heat flow from the flue gases to the boiler during dt. If C is the thermal capacitance of the boiler, then the energy balance equation is

$$C(dT) = \dot{m}_g c_{p,g}(T_{g,\text{in}} - T_{g,\text{out}})\,dt$$

The heat flow rate to the surface of the boiler can be expressed using the overall heat transfer coefficient and the logarithmic mean temperature difference (LMTD), $\Delta\overline{T}$, as

$$Q = UA\Delta\overline{T}$$

If T is the temperature of the water in the boiler, the LMTD is

$$\begin{aligned} \Delta\overline{T} &= \frac{(T_{g,\text{in}} - T) - (T_{g,\text{out}} - T)}{\ln\left(\frac{T_{g,\text{in}} - T}{T_{g,\text{out}} - T}\right)} \\ &= \frac{(1300°\text{F} - T) - (T_{g,\text{out}} - T)}{\ln\left(\frac{1300°\text{F} - T}{T_{g,\text{out}} - T}\right)} \end{aligned}$$

Equating this heat flow rate with the heat flow term on the right side of the energy balance equation gives

$$UA\left(\frac{(1300°\text{F} - T) - (T_{g,\text{out}} - T)}{\ln\left(\frac{1300°\text{F} - T}{T_{g,\text{out}} - T}\right)}\right) = \dot{m}_g c_{p,g}(T_{g,\text{in}} - T_{g,\text{out}})$$

Substituting known values,

$$\begin{aligned} &\left(5\ \frac{\text{Btu}}{\text{ft}^2\text{-hr-°F}}\right)(30{,}000\ \text{ft}^2) \\ &\quad\times\left(\frac{(1300°\text{F} - T) - (T_{g,\text{out}} - T)}{\ln\left(\frac{1300°\text{F} - T}{T_{g,\text{out}} - T}\right)}\right) \\ &= \left(100{,}000\ \frac{\text{lbm}}{\text{hr}}\right)\left(0.25\ \frac{\text{Btu}}{\text{lbm-°F}}\right)(1300°\text{F} - T_{g,\text{out}}) \end{aligned}$$

Solving for $T_{g,\text{out}}$,

$$\begin{aligned} \left(150{,}000\ \frac{\text{Btu}}{\text{hr-°F}}\right)(1300°\text{F} - T_{g,\text{out}}) &= \left(25{,}000\ \frac{\text{Btu}}{\text{hr-°F}}\right) \\ &\quad\times(1300°\text{F} - T_{g,\text{out}}) \\ &\quad\times\ln\left(\frac{1300°\text{F} - T}{T_{g,\text{out}} - T}\right) \\ 6 &= \ln\left(\frac{1300°\text{F} - T}{T_{g,\text{out}} - T}\right) \\ T_{g,\text{out}} &= (1300°\text{F} - T)e^{-6} + T \end{aligned}$$

Substituting this expression for $T_{g,\text{out}}$ into the energy balance equation gives

$$C(dT) = \dot{m}_g c_{p,g}\big(T_{g,\text{in}} - ((1300°\text{F} - T)e^{-6} + T)\big)\,dt$$

Substituting known values,

$$C(dT) = \left(100{,}000\ \frac{\text{lbm}}{\text{hr}}\right)\left(0.25\ \frac{\text{Btu}}{\text{lbm-°F}}\right) \times \left(1300°\text{F} - \left((1300°\text{F} - T)e^{-6} + T\right)\right)dt$$

Using the thermal capacitance found previously and simplifying,

$$\left(52{,}700\ \frac{\text{Btu}}{°\text{F}}\right)dT = \left(25{,}000\ \frac{\text{Btu}}{\text{hr-°F}}\right) \times (1297°\text{F} - 0.9975T)\,dt$$

The differential equation can now be integrated to find the time needed to heat the boiler from 90°F to 212°F. Separating variables,

$$\left(\frac{52{,}700\ \frac{\text{Btu}}{°\text{F}}}{1297°\text{F} - 0.9975T}\right)dT = \left(25{,}000\ \frac{\text{Btu}}{\text{hr-°F}}\right)dt$$

Integrating,

$$\int_{90°\text{F}}^{212°\text{F}} \left(\frac{52{,}700\ \frac{\text{Btu}}{°\text{F}}}{1297°\text{F} - 0.9975T}\right)dT = \int_0^{t_f} \left(25{,}000\ \frac{\text{Btu}}{\text{hr-°F}}\right)dt$$

$$52{,}832\ \frac{\text{Btu}}{°\text{F}} \int_{90°\text{F}}^{212°\text{F}} \left(\frac{1}{1300°\text{F} - T}\right) = \left(25{,}000\ \frac{\text{Btu}}{\text{hr-°F}}\right)t_f$$

$$\left(52{,}832\ \frac{\text{Btu}}{°\text{F}}\right)\left(-\ln(1300°\text{F} - T)\right)\Big|_{90°\text{F}}^{212°\text{F}} = \left(25{,}000\ \frac{\text{Btu}}{\text{hr-°F}}\right)t_f$$

$$\left(52{,}832\ \frac{\text{Btu}}{°\text{F}}\right)\left(-\ln(1300°\text{F} - 212°\text{F}) + \ln(1300°\text{F} - 90°\text{F})\right) = \left(25{,}000\ \frac{\text{Btu}}{\text{hr-°F}}\right)t_f$$

Solving for the time needed to heat the boiler,

$$t_f = (0.225\ \text{hr})\left(60\ \frac{\text{min}}{\text{hr}}\right) = 13.5\ \text{min} \quad (14\ \text{min})$$

The answer is (B).

Why Other Options Are Wrong

(A) This incorrect answer results when only the mass of the water is used to determine the thermal capacity.

(C) This incorrect answer results when T_1 is not subtracted from $T_{g,\text{in}}$ in the numerator.

(D) This incorrect answer results when the specific heat of steel is used instead of that of the flue gas.

SOLUTION 27

This problem can be solved with the correlation equation that uses the actual water loading, L_a, the measured air loading, G_a, the water loading at test temperatures, L_b, and the base air loading, G_b.

$$\frac{L_b}{L_a} = \left(\frac{G_b}{G_a}\right)^n$$

At constant conditions and using an average value of $n = 0.93$ the above equation becomes

$$\frac{G_b}{G_a} = \left(\frac{L_b}{L_a}\right)^{1/n} = \left(\frac{608\ \text{ft}^2}{576\ \text{ft}^2}\right)^{1/0.93} = 1.06$$

Therefore the air rate must be increased by 6%.

The answer is (C).

Why Other Options Are Wrong

(A) This incorrect answer results when 0.39 is used as the exponent instead of $1/0.93$.

(B) This incorrect answer results when 0.93 is used as the exponent instead of $1/0.93$.

(D) This incorrect answer results when $1/0.39$ is used as the exponent instead of $1/0.93$.

SOLUTION 28

Because the compressor is not cooled, the power necessary for isentropic compression of an ideal gas from p_1 to p_2 is known as the theoretical adiabatic power and is given by

$$P_{\text{th}} = \left(\frac{k}{k-1}\right)Q_1p_1\left(\left(\frac{p_2}{p_1}\right)^{\frac{k-1}{k}} - 1\right)$$

The ratio of specific heats is k, the flow rate is Q_1, the inlet pressure is p_1, and the discharge pressure is p_2. The ratio of specific heats is 1.4 for air, and the inlet pressure for the atmospheric air that enters this compressor is 14.7 psia. The adiabatic power is

$$P_{\text{th}} = \left(\frac{1.4}{1.4-1}\right)\left(50{,}000\ \frac{\text{ft}^3}{\text{min}}\right)\left(\frac{1\ \text{min}}{60\ \text{sec}}\right)\left(14.7\ \frac{\text{lbf}}{\text{in}^2}\right) \times \left(144\ \frac{\text{in}^2}{\text{ft}^2}\right)\left(\left(\frac{53\ \frac{\text{lbf}}{\text{in}^2}}{14.7\ \frac{\text{lbf}}{\text{in}^2}}\right)^{\frac{1.4-1}{1.4}} - 1\right) \times \left(\frac{1\ \text{hp}}{550\ \frac{\text{lbf-ft}}{\text{sec}}}\right) = 4968\ \text{hp}$$

The compressor is 79% efficient, so the motor shaft horsepower must be

$$\begin{aligned} P_{\text{shaft}} &= \frac{P_{\text{th}}}{\eta} \\ &= \frac{4968\ \text{hp}}{0.79} \\ &= 6289\ \text{hp} \quad (6300\ \text{hp}) \end{aligned}$$

The answer is (D).

Why Other Options Are Wrong

(A) This incorrect answer results when the flow is converted from cfm to cfs by dividing by $(60)^2$ instead of 60.

(B) This incorrect answer results when the specific heats term at the beginning of the adiabatic power equation is ignored.

(C) This incorrect answer results when 35 psia is used as the discharge pressure instead of 53 psia.

SOLUTION 29

The brake horsepower for the compressor can be determined using this graph, which gives the brake horsepower needed for compressing natural gas.

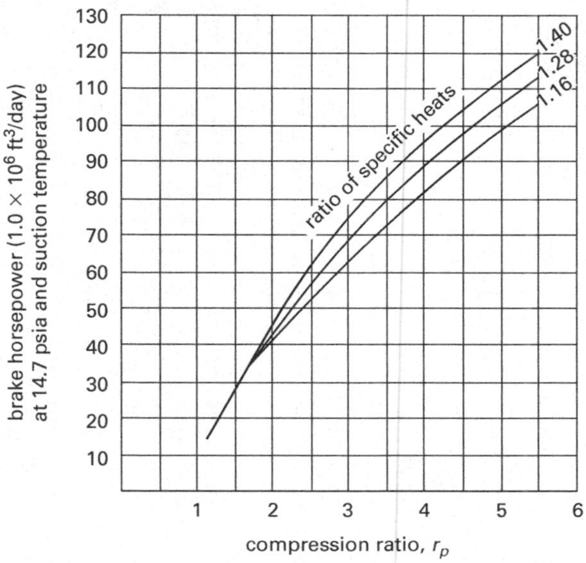

Reprinted from *Rules of Thumb for Mechanical Engineers*, Pope ed., p. 118, Copyright 1996, with permission from Elsevier.

The compression ratio, r_p, is calculated using the formula

$$r_p = \frac{p_d}{p_s}$$

The discharge pressure, p_d, and the suction pressure, p_s, must be absolute pressures. The atmospheric pressure at 3000 ft is 13.2 psia, so the compression ratio is

$$r_p = \frac{200\ \frac{\text{lbf}}{\text{in}^2} + 13.2\ \frac{\text{lbf}}{\text{in}^2}}{35\ \frac{\text{lbf}}{\text{in}^2} + 13.2\ \frac{\text{lbf}}{\text{in}^2}} = 4.42$$

From the graph, the brake horsepower required for $1.0 \times 10^6\ \text{ft}^3/\text{day}$ is approximately 95 hp. For $1.7 \times 10^6\ \text{ft}^3/\text{day}$, then, the horsepower requirement is

$$\begin{aligned} P &= (95\ \text{hp})\left(\frac{1.7 \times 10^6\ \frac{\text{ft}^3}{\text{day}}}{1.0 \times 10^6\ \frac{\text{ft}^3}{\text{day}}}\right) \\ &= 162\ \text{hp} \quad (160\ \text{hp}) \end{aligned}$$

The answer is (C).

Why Other Options Are Wrong

(A) This incorrect answer results when the correction factor for 1.7×10^6 ft^3/day is neglected.

(B) This incorrect answer results when absolute pressures are not used when the compression ratio is calculated, and the correction for 1.7×10^6 ft^3/day is neglected.

(D) This incorrect answer results when absolute pressures are not used when the compression ratio is calculated.

SOLUTION 30

Pump inlet pressure required is given as 20 psi. Convert this to feet of head at the given water temperature. The specific weight of water at 70°F is 62.3 lbf/ft^3.

$$\begin{aligned} h &= \frac{p}{\gamma} \\ &= \frac{\left(20\ \frac{\text{lbf}}{\text{in}^2}\right)\left(12\ \frac{\text{in}}{\text{ft}}\right)^2}{62.3\ \frac{\text{lbf}}{\text{ft}^3}} \\ &= 46.23\ \text{ft} \quad (46\ \text{ft}) \end{aligned}$$

The answer is (C).

Why Other Options Are Wrong

(A) This incorrect answer results when the inch to feet conversion is not squared.

(B) This incorrect answer results when the net positive suction head of 20 is given in psi instead of in feet of head.

(D) This incorrect answer results when the atmospheric pressure of 14.7 psi is added to the 20 psi.

SOLUTION 31

The torque can be found from the equation

$$\begin{aligned} P &= T\omega \\ T &= \frac{P}{\omega} \end{aligned}$$

P is the input power to the generator. This can be found by calculating the output power supplied by the generator and then using the generator's efficiency to calculate the necessary input power.

Find the output power. The angle ϕ is the power factor angle, and $\cos\phi$ is the power factor, p_f.

$$\begin{aligned} P_{\text{out}} &= \sqrt{3}\,VI\cos\phi \\ &= \sqrt{3}\,VI(\text{pf}) \\ &= \sqrt{3}\,(240\ \text{V})(20\ \text{A})(0.8) \\ &= 6650\ \text{W} \end{aligned}$$

Use the generator's efficiency to find the input power to the generator.

$$\begin{aligned} \eta &= \frac{P_{\text{out}}}{P_{\text{in}}} \\ P_{\text{in}} &= \frac{P_{\text{out}}}{\eta} = \frac{6650\ \text{W}}{0.90} \\ &= 7390\ \text{W} \end{aligned}$$

The torque needed to drive the generator at a rotational speed of 1800 rpm is

$$\begin{aligned} T &= \frac{P_{\text{in}}}{n} \\ &= \frac{\left(\dfrac{7390\ \text{W}}{1.356\ \dfrac{\text{W-sec}}{\text{ft-lbf}}}\right)}{\left(\dfrac{\left(1800\ \dfrac{\text{rev}}{\text{min}}\right)\left(2\pi\ \dfrac{\text{rad}}{\text{rev}}\right)}{60\ \dfrac{\text{sec}}{\text{min}}}\right)} \\ &= 28.91\ \text{ft-lbf} \quad (29\ \text{ft-lbf}) \end{aligned}$$

The answer is (B).

Why Other Options Are Wrong

(A) This incorrect answer results when power for a single-phase (without the square root of 3) is used when calculating P_{out}.

(C) This incorrect answer results when the power factor and efficiency numbers are interchanged.

(D) This incorrect answer results when P_{out} is calculated using 3 instead of the square root of 3.

SOLUTION 32

The lowest temperature that can be achieved using evaporative cooling occurs when the air leaves the cooler in its saturated state. Since evaporative cooling is almost identical to the adiabatic saturation process, the process follows lines of constant wet-bulb temperature on the psychrometric chart. From the psychrometric chart, the minimum temperature is found to be 71°F.

The answer is (C).

Why Other Options Are Wrong

(A) This incorrect option is the enthalpy value.

(B) This incorrect answer results when a horizontal line on the psychrometric chart is followed to the left until it intersects the saturation line.

(D) This incorrect answer results when the specific volume line from the bottom of the psychrometric chart at 115°F is followed until it nearly intersects the saturation line.

SOLUTION 33

The pump should be run at the speed that will provide the highest efficiency. From the figure "Average Pump Efficiency Versus Specific Speed" in *Mechanical Engineering Reference Manual*, the maximum efficiency for a pump delivering 3000 gpm is approximately 87%, and it occurs at a specific speed, n_s, of about 2500. Use the equation for specific speed, and solve for the pump speed, n.

$$n_s = \frac{n\sqrt{Q_{\text{gpm}}}}{h_{A,\text{ft}}^{0.75}}$$

$$\begin{aligned} n &= \frac{n_s h_{A,\text{ft}}^{0.75}}{\sqrt{Q_{\text{gpm}}}} \\ &= \frac{\left(2500\ \frac{\text{rev}}{\text{min}}\right)(100\ \text{ft})^{0.75}}{\sqrt{3000\ \frac{\text{gal}}{\text{min}}}} \\ &= 1440\ \text{rev/min} \quad (1400\ \text{rpm}) \end{aligned}$$

The answer is (B).

Why Other Options Are Wrong

(A) This incorrect answer results when an incorrect specific speed, $n_s = 2000$, is used.

(C) This incorrect answer results when the flow rate of 3000 gpm is converted to cubic feet per minute.

(D) This incorrect answer results from switching n_s and n in the equation for specific speed.

SOLUTION 34

The change in entropy across a normal shock wave can be found using the equation

$$\Delta S_{x\to y} = c_p \ln\frac{T_y}{T_x} - R\ln\frac{p_y}{p_x}$$

The subscripts x and y represent the upstream and downstream conditions of the shock wave, respectively. From a table of the properties of air at standard temperature and pressure, the specific heat at constant pressure, c_p, is 0.24 Btu/lbm-°R, and the specific gas constant, R, is 53.35 ft-lbf/lbm-°R. From a table of normal shock parameters, for a Mach number of 2.2,

$$\frac{T_y}{T_x} = 1.857$$

$$\frac{p_y}{p_x} = 5.480$$

The change in entropy is

$$\begin{aligned} \Delta S_{x\to y} &= \left(0.24\ \frac{\text{Btu}}{\text{lbm-°R}}\right)\ln 1.857 \\ &\quad - \left(\frac{53.35\ \frac{\text{ft-lbf}}{\text{lbm-°R}}}{778\ \frac{\text{ft-lbf}}{\text{Btu}}}\right)\ln 5.480 \\ &= 0.03190\ \text{Btu/lbm-°R} \quad (0.032\ \text{Btu/lbm-°R}) \end{aligned}$$

The answer is (A).

Why Other Options Are Wrong

(B) This incorrect answer results when the second term in the equation for change in entropy is added to the first term instead of subtracted from it.

(C) This incorrect answer results when the ratio $p_{0,y}/p_{0,x}$ is used instead of p_y/p_x.

(D) This incorrect answer results when the T_y/T_x and p_y/p_x values are interchanged.

SOLUTION 35

This is Fanno flow because the system is adiabatic (the duct is insulated), the duct has a constant cross-sectional area, there is no work being done, and fluid friction is present.

The ratio of specific heats for helium is 1.67. From a Fanno flow gas dynamics table for $k = 1.67$, the T/T^* ratio at M = 0.1 is 1.335 (1.3).

If a table is not available, then the following equation for Fanno flow can be used.

$$\frac{T}{T^*} = \frac{\frac{k+1}{2}}{1+\left(\frac{k-1}{2}\right)\text{M}^2} = \frac{\frac{1.67+1}{2}}{1+\left(\frac{1.67-1}{2}\right)(0.1)^2} = 1.331 \quad (1.3)$$

The answer is (C).

Why Other Options Are Wrong

(A) This incorrect answer results when a Rayleigh flow table is used.

(B) This incorrect answer results when the T/T_0 column in an isentropic flow table is used.

(D) This incorrect answer results when the division by 2 in the numerator and denominator is neglected.

SOLUTION 36

The change in pressure can be calculated using

$$\Delta p = f\left(\frac{L}{d}\right)\left(\frac{\rho \text{v}^2}{2}\right)$$

The pressure drop per unit length can therefore be expressed as

$$\frac{\Delta P}{L} = \frac{f\rho \text{v}^2}{2d}$$

The friction factor, f, and the density of oil, ρ, are the same for the feeder lines and the main line. Velocity can be expressed as

$$\text{v} = \frac{Q}{A} = \frac{Q}{\frac{\pi d^2}{4}} = \frac{4Q}{\pi d^2}$$

If Q is the flow rate in each of the feeder lines, the flow rate in the main line is $3Q$. From a table of pipe dimensions, 16 in schedule-40 pipe has an internal diameter of 15 in, and 30 in schedule-40 pipe has an internal diameter of 29 in. The ratio of the pressure drop per unit length in the main line to that in the feeder lines is therefore

$$\frac{\left(\frac{\Delta p}{L}\right)_{\text{main}}}{\left(\frac{\Delta p}{L}\right)_{\text{feed}}} = \frac{\frac{f\rho \text{v}^2_{\text{main}}}{2d_{\text{main}}}}{\frac{f\rho \text{v}^2_{\text{feed}}}{2d_{\text{feed}}}} = \frac{\frac{\text{v}^2_{\text{main}}}{d_{\text{main}}}}{\frac{\text{v}^2_{\text{feed}}}{d_{\text{feed}}}} = \frac{\frac{\left(\frac{4(3Q)}{\pi d^2_{\text{main}}}\right)^2}{d_{\text{main}}}}{\frac{\left(\frac{4Q}{\pi d^2_{\text{feed}}}\right)^2}{d_{\text{feed}}}} = \frac{9d^5_{\text{feed}}}{d^5_{\text{main}}} = \frac{(9)(15 \text{ in})^5}{(29 \text{ in})^5} = 0.3332 \quad (0.33)$$

The answer is (B).

Why Other Options Are Wrong

(A) This incorrect answer results when the flow in the main line is taken as Q instead of $3Q$.

(C) This incorrect answer results when the velocities are not squared.

(D) This incorrect answer is the ratio of the feeder line to the main line.

3 Energy/Power System Applications

PROBLEM 1

The number of independent quantities needed to specify the thermodynamic state of a system in equilibrium is equal to

(A) three

(B) the number of phases of the substance present in the system

(C) the number of possible phases a substance can have

(D) the number of possible work modes plus one

Hint: The number of independent quantities is not a fixed number.

PROBLEM 2

Air at a dry-bulb temperature of 84°F has a relative humidity of 50%. The heat needed to heat 2500 lbm/hr of this air to 102°F without changing the moisture content is most nearly

(A) 6.5×10^3 Btu/hr

(B) 11×10^3 Btu/hr

(C) 23×10^3 Btu/hr

(D) 38×10^3 Btu/hr

Hint: A psychrometric chart will be useful.

PROBLEM 3

An unknown amount of C_3H_8 (propane) is burned in the air. An Orsat (dry product) analysis gives results of 6.6% O_2, 7.7% CO_2, and 2.2% CO. The excess air in the reaction is most nearly

(A) 17%

(B) 25%

(C) 33%

(D) 90%

Hint: Determine the stoichiometric reaction.

PROBLEM 4

Consider two Brayton cycles. In both systems, there is isentropic compression in the compressor (A-B) and constant-pressure heat addition in the combustor (B-C).

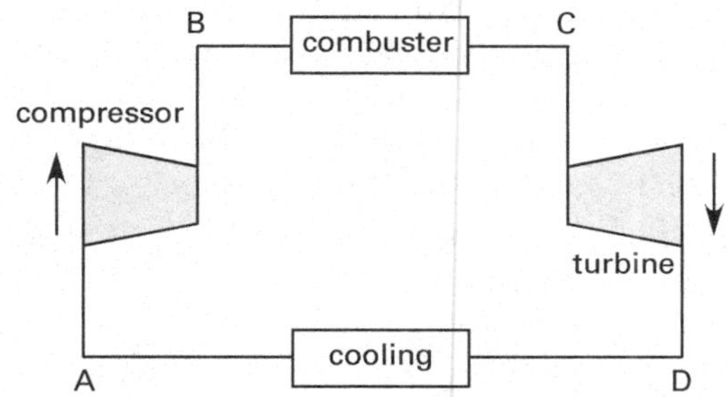

In system 1, an isentropic expansion occurs in the turbine (C-D), followed by constant-pressure cooling (D-A). In system 2, the turbine isentropic efficiency is 75%. The two systems operate such that points A, B, and C are the same on a *T-s* diagram for each. Which of the following statements are true?

I. The thermal efficiency of system 1 is higher than that of system 2.

II. The work of the compressor is the same in systems 1 and 2.

III. More heat is rejected during cooling in system 1 than in system 2.

IV. The work out of the turbine is greater in system 1 than in system 2.

(A) I and II only

(B) III and IV only

(C) I, II, and IV only

(D) I, II, III, and IV

Hint: No calculations are required.

PROBLEM 5

A combustion furnace uses 2 lbmol of H_2 and 1 lbmol of O_2 at 537°R and produces 2 lbmol of water vapor in the exhaust. If the products have an average specific heat of

1.15 Btu/lbm-°R, the maximum possible exit temperature is most nearly

(A) 1200°R

(B) 1800°R

(C) 5000°R

(D) 5500°R

Hint: The maximum temperature is equal to the adiabatic flame temperature.

PROBLEM 6

If the steam generated in a process has a circulation ratio of five, which of the following statements is true?

(A) The steam-water mixture contains five times more water than steam.

(B) The steam-water mixture contains five times more steam than water.

(C) 1 lbm of the steam-water mixture contains 0.2 lbm of water.

(D) 1 lbm of the steam-water mixture contains 0.2 lbm of steam.

Hint: Think about steam quality.

PROBLEM 7

A preliminary design to transport 1000 gpm of water from a reservoir to an open tank is shown in *Illustration for Problem 7.* The line includes four shutoff gate valves, one swing check valve, and two 45° elbows. If all the fittings are flange fittings, the pump's total dynamic head is most nearly

(A) 140 ft

(B) 150 ft

(C) 180 ft

(D) 210 ft

Hint: The friction head loss for each section of pipe must be calculated separately.

PROBLEM 8

Part of the repair to a center pipe-to-flange connection in a reactor is shown. The main exit pipe (not shown) was placed over the short piece of pipe labeled in the drawing. These two pipes were then welded together to complete the repair. The connection is held by 38 half-inch UNF 20 bolts. The bolts are ASTM A354 grade BC. The construction is such that the bolts are subjected to a purely axial load and are designed to fail before the weld.

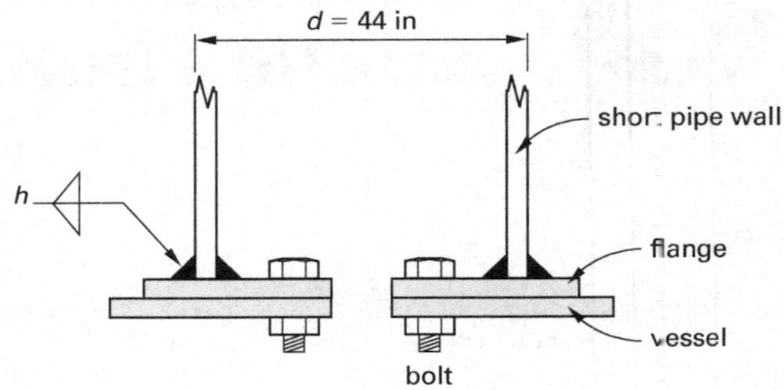

cross section of pipe

(not to scale)

If the weld is to be twice as strong as the bolts and if the allowable shear stress in the weld is 35 ksi, the minimum weld depth is most nearly

(A) 0.250 in

(B) 0.500 in

(C) 0.625 in

(D) 0.750 in

Hint: Pay attention to how the external loading in a fillet weld is carried.

PROBLEM 9

The block diagram for the control of an electrically actuated hydraulic valve is shown.

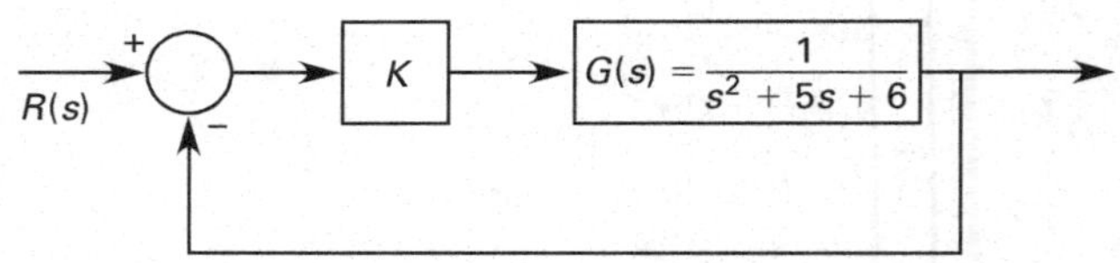

The largest value of the gain, K, that will maintain a stable, closed-loop system is most nearly

(A) 0.25

(B) 2.5

(C) 6.0

(D) ∞

Hint: The open-loop transfer function is important.

Illustration for Problem 7

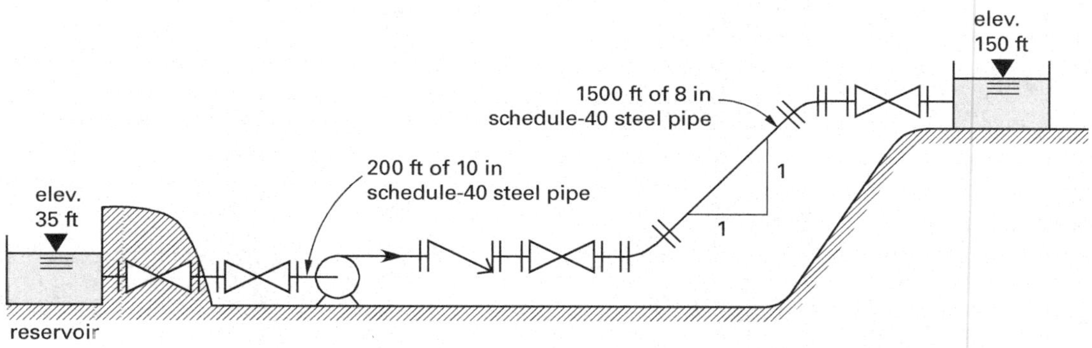

Energy/Power System Apps.

PROBLEM 10

A one-pass, crossflow radiator uses water as the engine coolant and has the following characteristics.

- core height: 255 mm
- core width: 330 mm
- core depth: 26 mm
- coefficient of heat transfer: 62.9 W/m^2·°C
- air heat transfer surface area: 3.1 m^2
- water heat transfer surface area: 0.37 m^2

The radiator is being tested under the following conditions.

- water flow rate: 20 L/min
- air velocity at the radiator face: 3 m/s
- inlet air temperature: 40°C
- outlet air temperature: 67°C
- inlet water temperature: 100°C
- outlet water temperature: 94°C

The heat transfer rate is most nearly

(A) 1.0 kW

(B) 8.3 kW

(C) 18 kW

(D) 37 kW

Hint: Use a counterflow heat exchanger model.

PROBLEM 11

The throttling area in a needle valve is a function of the

I. seat opening diameter

II. needle lift

III. cone height

IV. cone half angle

(A) I and IV

(B) II and III

(C) I, II, and IV

(D) I, III, and IV

Hint: The throttling areas for a needle valve and a poppet valve are calculated using the same equation.

PROBLEM 12

A manufacturer of electrically actuated four-way hydraulic valves uses the following equation to compute the flow through its valves.

$$Q_R = KI_m\sqrt{P_v}$$

Q_R is the rated flow in gallons per minute, K is a valve constant, I_m is the maximum DC current supplied to the valve (15 milliamps), and P_v is the pressure drop across the valve. The manufacturer rates its valves assuming a 1000 psi drop across the valve, and has 1, 5, 10, and 15 gpm valves available, cost increasing with size. In a particular application, the supply and peak load pressures are 1500 psi and 1100 psi respectively. The least expensive valve that will supply a 23 in^3/sec flow under these conditions is the

(A) 1 gpm valve

(B) 5 gpm valve

(C) 10 gpm valve

(D) 15 gpm valve

Hint: The valve constant is key here.

PROBLEM 13

The supplementary firing system of a heat recovery steam generator burns natural gas in the flue gas with a thermal efficiency of 55%. The higher heating value of the natural gas used is 17,000 Btu/lbm. The mass flow rate of the flue gas through the firing system is 3,180,000 lbm/hr. The combustion process in the firing system occurs at constant pressure, and the specific heat of the flue gas before and after combustions is approximately 0.248 Btu/lbm-°F. The natural gas is fed into the firing system at a temperature of 60°F and a mass flow rate of 65,000 lbm/hr. The specific heat of the natural gas is 0.692 Btu/lbm-°F. If the flue gas enters the firing system at a temperature of 980°F, its temperature as it leaves is most nearly

(A) 700°F

(B) 1700°F

(C) 1800°F

(D) 2100°F

Hint: Start by finding the thermal energy provided by the combustion of the natural gas.

PROBLEM 14

The ideal vapor refrigeration cycle shown uses refrigerant-134a and operates between 20 psi and 120 psi.

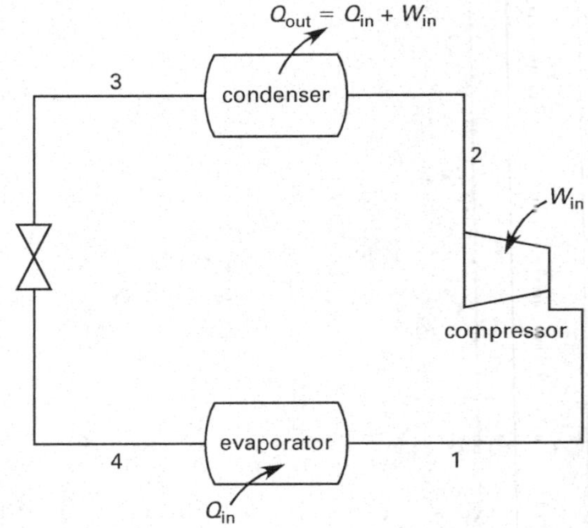

state 1: saturated vapor

$P_1 = 20$ psi

$h_1 = h_{g@20\text{ psi}} = 102.73$ Btu/lbm

$s_1 = s_{g@20\text{ psi}} = 0.22567$ Btu/lbm-°R

state 2: superheated vapor

$P_2 = 120$ psi

$s_2 = s_1 = 0.22567$ Btu/lbm-°R

state 3: saturated liquid

$P_3 = 120$ psi

$h_3 = h_{f@120\text{ psi}} = 41.787$ Btu/lbm

state 4: low-quality saturated mixture

$h_4 \approx h_3$ (throttling) $= 41.787$ Btu/lbm

The system is to produce a cooling effect of 5 tons and operate with a coefficient of performance equal to 3.8. The compressor horsepower needed is most nearly

(A) 4.6 hp

(B) 6.2 hp

(C) 14 hp

(D) 30 hp

Hint: Use the COP equation to eliminate a table lookup.

PROBLEM 15

An underwater mooring system designed to restrict roll motion consists of two rings that are each 1 ft in diameter, spaced 4 ft apart, and attached by crisscrossed 1/16 in stainless steel 1×19 strand cables. The fill factor for 1×19 strand cable is 0.76.

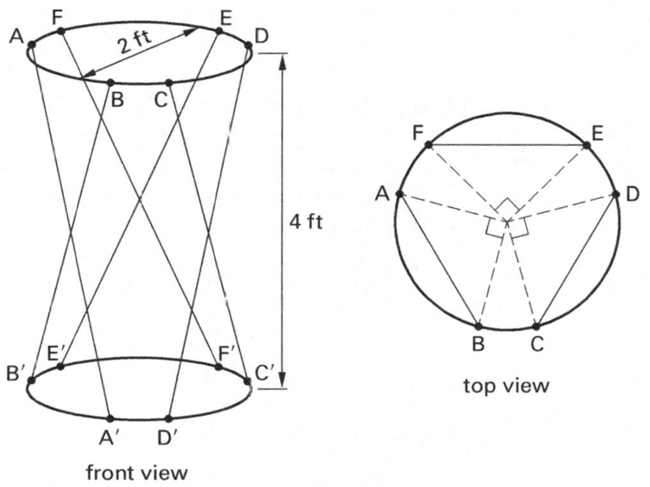

The maximum torque-resisting capacity of a mooring bay is 1800 in-lbf. The normal stress in the cables at the maximum torque is most nearly

(A) 43 ksi

(B) 45 ksi

(C) 69 ksi

(D) 91 ksi

Hint: Examine one pair of cables.

PROBLEM 16

The beam shown is made from 1018 steel and has a completely reversed, repeated load P applied to it.

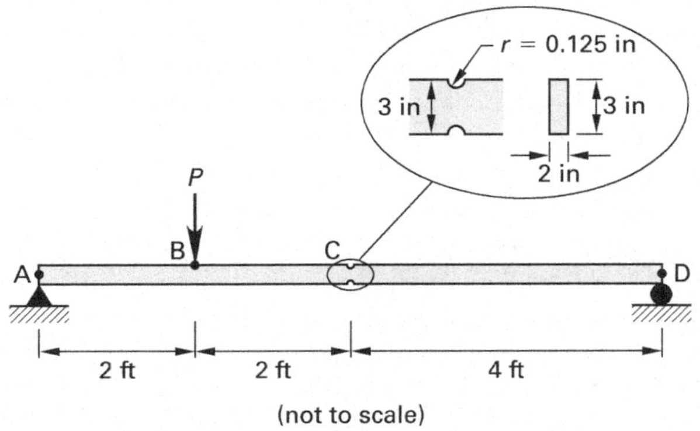

Use the modified Goodman fatigue failure criterion and a factor of safety of three. All endurance limit modifying factors are equal to one. The maximum allowable value for P is most nearly

(A) 730 lbf

(B) 870 lbf

(C) 1400 lbf

(D) 1800 lbf

Hint: Find the critical point.

PROBLEM 17

The topping cycle of a combined cycle is a gas turbine cycle that operates with a pressure ratio equal to 8. Air enters the turbine at 2340°R and the compressor at 540°R. The isentropic efficiencies of the gas turbine and the compressor are 85% and 80%, respectively. The bottoming cycle is an ideal Rankine cycle operating between 1000 psi and 0.75 psi. Steam is heated to 930°F in a heat exchanger by the exhaust gases that leave the heat exchanger at 810°R.

An analysis of the gas turbine cycle shows that the heat transfer per unit mass is 340 Btu/lbm, the net work done per unit mass is 90 Btu/lbm, and the efficiency is 27%. An analysis of the heat exchanger shows that only 0.131 lbm of steam can be heated from 90°F to 930°F using 1 lbm of exhaust gases as the gases cool from 1535°R to 810°R. An analysis of the steam cycle shows that the net work done per unit mass is 570 Btu/lbm with an efficiency of 41%. The themal efficiency of this combined cycle is most nearly

(A) 11%

(B) 22%

(C) 48%

(D) 83%

Hint: The ratio of the steam mass-flow rate to the gas turbine mass-flow rate will be needed.

Energy/Power System Apps.

PROBLEM 18

A double-acting 3.0 inch pneumatic cylinder with a 0.125 in diameter piston rod operates with a supply pressure of 100 psi. Assuming a typical efficiency of 0.85 to account for cylinder friction, the thrust generated on the return stroke is most nearly

(A) 200 lbf

(B) 300 lbf

(C) 600 lbf

(D) 700 lbf

Hint: Be sure to know the difference in geometry between the forward and return strokes.

SOLUTION 1

To specify the state of a substance, an intensive property such as pressure, electrical potential, or magnetic potential is needed for each work mode, in addition to an intensive property to describe heat interactions.

The answer is (D).

Why Other Options Are Wrong

(A) This answer is incorrect because the number of independent quantities needed varies depending on the substance in the system.

(B) This answer is incorrect because the number of phases of the substance is irrelevant.

(C) This answer is incorrect because the number of phases of the substance is irrelevant.

SOLUTION 2

The heat needed can be calculated using the formula

$$Q = \dot{m}\Delta h$$

From a psychrometric chart, the enthalpy, h, for 84°F air with 50% relative humidity is

$$h_{84} = 34.1 \text{ Btu/lbm}$$

The 102°F air must have the same humidity ratio as the 84°F air, because the moisture content is to remain unchanged. Again from the psychrometric chart, the humidity ratio is

$$\omega = 0.0125$$

From the psychrometric chart, the enthalpy for 102°F air with a 0.0125 humidity ratio is

$$h_{102} = 38.6 \text{ Btu/lbm}$$

The heat needed is

$$\begin{aligned} Q = \dot{m}\Delta h &= \left(2500\ \frac{\text{lbm}}{\text{hr}}\right)\left(38.6\ \frac{\text{Btu}}{\text{lbm}} - 34.1\ \frac{\text{Btu}}{\text{lbm}}\right) \\ &= 11{,}250 \text{ Btu/hr} \quad (11 \times 10^3 \text{ Btu/hr}) \end{aligned}$$

The answer is (B).

Why Other Options Are Wrong

(A) This incorrect answer results when the 60% relative humidity line is used instead of the 50% relative humidity line.

(C) This incorrect answer results when a mass flow rate of 5200 lbm/hr, instead of 2500, is used.

(D) This incorrect answer results from incorrectly assuming that the relative humidity for both temperatures must be equal.

SOLUTION 3

The balanced stoichiometric reaction must be found first. For this stoichiometric reaction, only H_2O, CO_2, and N_2 are allowed as products. Thus, for 1 mol of propane,

$$C_3H_8 + a(O_2 + 3.773N_2) \rightarrow bH_2O + cCO_2 + dN_2$$

From conservation of mass and the three carbon atoms on the left side, there must be three CO_2 molecules on the right side, and therefore c equals 3. There are eight H atoms on the left, so there must be four H_2O molecules on the right, and b equals 4. There are then 10 atoms of oxygen on the right side, so there must be five O_2 molecules on the left, and a equals 5. Finally, there are 5×3.773 or 18.87 molecules of N_2 per mole of propane on the left, so there must be 18.87 molecules of N_2 on the right, and d equals 18.87. The balanced, stoichiometric equation is then

$$C_3H_8 + 5(O_2 + 3.773N_2) \rightarrow 4H_2O + 3CO_2 + 18.87N_2$$

Next, analyze the Orsat data to determine the actual reaction ratios. The reaction with the extended product list is

$$\begin{aligned} &eC_3H_8 + f(O_2 + 3.773N_2) \\ &\quad \rightarrow gH_2O + hCO_2 + iO_2 + jCO + kN_2 \end{aligned}$$

For every 100 mol of Orsat products, there are 7.7 mol of CO_2, 6.6 mol of O_2, and 2.2 mol of CO. The remaining 83.5 mol must be N_2. (The Orsat analysis is a dry analysis and does not provide information about the water content.) Substituting,

$$\begin{aligned} &eC_3H_8 + f(O_2 + 3.773N_2) \\ &\quad \rightarrow gH_2O + 7.7CO_2 + 6.6O_2 + 2.2CO + 83.5N_2 \end{aligned}$$

Conservation of mass can be used to determine the remaining unknowns. For carbon,

$$\begin{aligned} 3e &= 7.7 + 2.2 \\ e &= 3.3 \end{aligned}$$

Then, for hydrogen,

$$\begin{aligned} (8)(3.3) &= 2g \\ g &= 13.2 \end{aligned}$$

And then, for oxygen,

$$\begin{aligned} 2f &= 13.2 + 15.4 + 13.2 + 2.2 \\ f &= 22 \end{aligned}$$

Check the nitrogen balance. On the left are 166 nitrogen atoms. On the right are 167. Taking rounding errors into account, particularly with the Orsat analysis numbers, this is a good balance.

The final reaction is then

$$\begin{aligned} &3.3C_3H_8 + 22(O_2 + 3.773N_2) \\ &\quad \rightarrow 13.2H_2O + 7.7CO_2 + 6.6O_2 + 2.2CO + 83.5N_2 \end{aligned}$$

In the stoichiometric reaction, there are 16.5 mol of air for every 3.3 mol of propane. There are 3.3 mol of propane and 22 mol of air in the actual reaction, so the excess air is

$$\begin{aligned} \text{excess air} &= \frac{\text{actual air-stoichiometric air}}{\text{stoichiometric air}} \\ &= \frac{22 \text{ mol} - 16.5 \text{ mol}}{16.5 \text{ mol}} \\ &= 0.33 \quad (33\%) \end{aligned}$$

The answer is (C).

Why Other Options Are Wrong

(A) This incorrect answer results when the CO and CO_2 values are reversed in the Orsat analysis.

(B) This incorrect answer results when the excess air is calculated relative to the actual air instead of to the stoichiometric air.

(D) This incorrect answer results when the stoichiometric reaction is mistakenly calculated with CO instead of CO_2.

SOLUTION 4

Plot both cycles on the same T-s diagram.

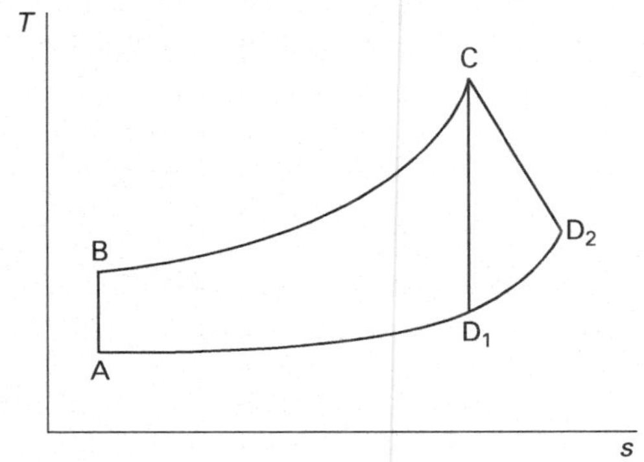

Energy/Power System Apps.

Thermal efficiency is defined as

$$\eta_{th} = \frac{W_{turbine} - W_{compressor}}{Q_{in}}$$

Q_{in} depends on T_C and T_B, and $W_{compressor}$ depends on T_A and T_B, so both values are the same for both cycles. $W_{turbine}$, however, is larger for system 1 (related to $T_D - T_C$), so the thermal efficiency of system 1 is higher, and statement I is true.

As just described, the work of the two compressors is the same, so statement II is true.

The heat rejected during cooling varies in proportion to $T_D - T_A$, which is larger for system 2. Thus, statement III is false.

The work out of the turbine varies in proportion to $T_C - T_D$, which is larger for system 1. Thus, statement IV is true.

The answer is (C).

Why Other Options Are Wrong

(A) This choice is wrong because statement IV is also true.

(B) This choice is wrong because statements I and II are true, and III is false.

(D) This choice is wrong because statement III is false.

SOLUTION 5

To estimate the maximum temperature, determine the adiabatic flame temperature.

$$T_{max} = T_i + \frac{\text{lower heat of combustion}}{m_{products} c_{p,mean}}$$

The initial temperature, T_i, is 537°R.

Water vapor has a molecular weight of 18 lbm/lbmol. The mass of 2 lbmol of water vapor is 36 lbm.

The mass must be normalized by the mass of the fuel, to correspond to the normalization of the lower heat of combustion.

$$m_{products} = \frac{m_{wv}}{m_{H_2}} = \frac{36 \text{ lbm}}{4 \text{ lbm}_{H_2}} = 9 \text{ lbm/lbm}_{H_2}$$

The lower heat of combustion is 51,623 Btu/lbm$_{H_2}$.

The maximum temperature can now be found.

$$\begin{aligned} T_{max} &= T_i + \frac{\text{lower heat of combustion}}{m_{products} c_{p,mean}} \\ &= 537°\text{R} + \frac{51{,}623 \frac{\text{Btu}}{\text{lbm}_{H_2}}}{\left(9 \frac{\text{lbm}}{\text{lbm}_{H_2}}\right)\left(1.15 \frac{\text{Btu}}{\text{lbm-°R}}\right)} \\ &= 5520°\text{R} \quad (5500°\text{R}) \end{aligned}$$

The answer is (D).

Why Other Options Are Wrong

(A) This incorrect answer results when the product mass is not normalized and the initial temperature is omitted.

(B) This incorrect answer results when the product mass is not normalized by the mass of the fuel.

(C) This incorrect answer results when the initial temperature is omitted.

SOLUTION 6

Using x as the steam quality, the circulation ratio, or CR, is defined as

$$\text{CR} = \frac{1}{x}$$

A circulation ratio of five gives a steam quality of 0.2, or 20%, meaning that 1 lbm of the steam-water mixture contains 0.2 lbm of steam.

The answer is (D).

Why Other Options Are Wrong

(A) Using the CR definition reveals that there is four times more water in the mixture than steam.

(B) Using the CR definition reveals that there is four times more water in the mixture than steam.

(C) The quality factor measures the amount of dry steam, not the amount of water in a steam-water mixture.

SOLUTION 7

At least three solution methods are possible.

Method 1

The pump's total dynamic head or pump head can be calculated using the Bernoulli equation.

$$\frac{p_1}{\gamma} + \frac{v_1^2}{2g} + z_1 + h_A = \frac{p_2}{\gamma} + \frac{v_2^2}{2g} + z_2 + h_f$$

With the reservoir and tank surfaces as points 1 and 2, respectively, the gauge pressures and velocities at the surfaces are zero.

$$p_1 = v_1 = p_2 = v_2 = 0$$

The Bernoulli equation now becomes

$$\begin{aligned} 0 + 0 + z_1 + h_A &= 0 + 0 + z_2 + h_f \\ 35\text{ ft} + h_A &= 150\text{ ft} + h_f \end{aligned}$$

h_A is the pump's total dynamic head (TDH).

The head loss due to friction, h_f, can be found using the method of equivalent lengths. Since the suction and discharge lines are of different diameters, the friction head loss for each section must be calculated separately and then added together to find the total friction head loss.

Suction

The equivalent length for a 10 in, flanged, schedule-40, steel gate valve is 3.2 ft. The equivalent suction line length is

$$L_{\text{eq}} = 200\text{ ft} + (2)(3.2\text{ ft}) = 206.4\text{ ft}$$

The head loss per 100 linear feet of pipe, h_f, can be found using pipe friction tables. Be sure to use the water, schedule-40 steel pipe table. Looking up h_f for a 10 in pipe with a flow rate of 1000 gpm, it is found that

$$h_f = \frac{0.5\text{ ft}}{100\text{ ft}}$$

The suction line head loss is

$$h_{f,S} = \left(\frac{0.5\text{ ft}}{100\text{ ft}}\right)(206.4\text{ ft}) = 1.03\text{ ft}$$

Discharge

The equivalent length for an 8 in, flanged, schedule-40, steel gate valve is 3.2 ft. The equivalent length for an 8 in, flanged, schedule-40, steel swing valve is 90 ft. The equivalent length for an 8 in, flanged, schedule-40, regular steel 45° elbow is 7.7 ft. The equivalent discharge line length is

$$\begin{aligned} L_{\text{eq}} &= 1500\text{ ft} + (2)(3.2\text{ ft}) + 90\text{ ft} + (2)(7.7\text{ ft}) \\ &= 1612\text{ ft} \end{aligned}$$

Using the pipe friction tables, h_f is

$$h_f = \frac{1.56\text{ ft}}{100\text{ ft}}$$

The discharge line head loss is

$$h_{f,D} = \left(\frac{1.56\text{ ft}}{100\text{ ft}}\right)(1612\text{ ft}) = 25.1\text{ ft}$$

The total head loss is

$$h_f = h_{f,S} + h_{f,D} = 1.03\text{ ft} + 25.1\text{ ft} = 26.1\text{ ft}$$

The pump's TDH can now be found.

$$\begin{aligned} 35\text{ ft} + h_A &= 150\text{ ft} + 26.1\text{ ft} \\ h_A &= 141.1\text{ ft} \quad (140\text{ ft}) \end{aligned}$$

The answer is (A).

Method 2

The method of equivalent lengths is still used, but in this method the Darcy equation and the Moody friction factor, f, are used to determine the friction head losses. The Darcy equation is

$$h_f = \frac{fL_{\text{eq}}v^2}{2dg}$$

The suction and discharge lines are of different diameters, so the friction head loss for each section must be calculated separately and then added together to find the total friction head loss.

Suction

The velocity can be calculated from the flow rate using the formula

$$Q = vA$$

$$\left(1000\ \frac{\text{gal}}{\text{min}}\right)\left(\frac{1\text{ min}}{60\text{ sec}}\right) \times \left(\frac{1\text{ ft}^3}{7.48\text{ gal}}\right) = v\left(\frac{\pi(10.02\text{ in})^2}{4}\right)\left(\frac{1\text{ ft}}{12\text{ in}}\right)^2$$

$$v = 4.07\text{ ft/sec}$$

The Reynolds number is needed to find the Moody friction factor.

$$\begin{aligned} \text{Re} = \frac{vd}{\nu} &= \frac{\left(4.07\ \frac{\text{ft}}{\text{sec}}\right)(10.02\text{ in})\left(\frac{1\text{ ft}}{12\text{ in}}\right)}{1.4 \times 10^{-5}\ \frac{\text{ft}^2}{\text{sec}}} \\ &= 2.4 \times 10^5 \end{aligned}$$

The relative roughness is also needed (the specific roughness, ϵ, for steel is 0.0002 ft).

$$\frac{\epsilon}{d} = \frac{0.0002 \text{ ft}}{(10.02 \text{ in})\left(\frac{1 \text{ ft}}{12 \text{ in}}\right)} = 0.00024$$

Using the Moody friction factor chart, the friction factor is

$$f = 0.017$$

The friction head loss in the suction line is

$$h_{f,S} = \frac{(0.017)(206.4 \text{ ft})\left(4.07 \frac{\text{ft}}{\text{sec}}\right)^2}{(2)(10.02 \text{ in})\left(\frac{1 \text{ ft}}{12 \text{ in}}\right)\left(32.2 \frac{\text{ft}}{\text{sec}^2}\right)} = 1.1 \text{ ft}$$

Discharge

The velocity for the discharge line is found from the equation

$$Q = \text{v}A$$

$$\left(1000 \frac{\text{gal}}{\text{min}}\right)\left(\frac{1 \text{ min}}{60 \text{ sec}}\right) \times \left(\frac{1 \text{ ft}^3}{7.48 \text{ gal}}\right) = \text{v}\left(\frac{\pi(7.981 \text{ in})^2}{4}\right)\left(\frac{1 \text{ ft}}{12 \text{ in}}\right)^2$$

$$\text{v} = 6.41 \text{ ft/sec}$$

The Reynolds number is

$$\text{Re} = \frac{\text{v}d}{\nu} = \frac{\left(6.41 \frac{\text{ft}}{\text{sec}}\right)(7.981 \text{ in})\left(\frac{1 \text{ ft}}{12 \text{ in}}\right)}{1.4 \times 10^{-5} \frac{\text{ft}^2}{\text{sec}}} = 3.0 \times 10^5$$

The relative roughness is

$$\frac{\epsilon}{d} = \frac{0.0002 \text{ ft}}{(7.981 \text{ in})\left(\frac{1 \text{ ft}}{12 \text{ in}}\right)} = 0.0003$$

From the Moody friction factor chart, the friction factor is

$$f = 0.017$$

The friction head loss in the discharge line is

$$h_{f,D} = \frac{(0.017)(1612 \text{ ft})\left(6.41 \frac{\text{ft}}{\text{sec}}\right)^2}{(2)(7.981 \text{ in})\left(\frac{1 \text{ ft}}{12 \text{ in}}\right)\left(32.2 \frac{\text{ft}}{\text{sec}^2}\right)} = 26.3 \text{ ft}$$

The total head loss is

$$\begin{aligned} h_f &= h_{f,S} + h_{f,D} \\ &= 1.1 \text{ ft} + 26.3 \text{ ft} \\ &= 27.4 \text{ ft} \end{aligned}$$

The pump's TDH is

$$\begin{aligned} 35 \text{ ft} + h_A &= 150 \text{ ft} + 27.4 \text{ ft} \\ h_A &= 142.4 \text{ ft} \quad (140 \text{ ft}) \end{aligned}$$

It should be noted that the Darcy friction factor table could also be used for f. Using a Darcy friction factor table, f is 0.017 for both the suction and discharge lines.

The answer is (A).

Method 3

If the pipe lengths are long and there are few fittings, the losses due to the fittings can be neglected to provide an approximation that can be used to select one of the answers.

The suction line head loss is

$$h_{f,S} = \left(\frac{0.5 \text{ ft}}{100 \text{ ft}}\right)(200 \text{ ft}) = 1.0 \text{ ft}$$

The discharge-line head loss is

$$h_{f,D} = \left(\frac{1.56 \text{ ft}}{100 \text{ ft}}\right)(1500 \text{ ft}) = 23.4 \text{ ft}$$

The pump's TDH is

$$\begin{aligned} 35 \text{ ft} + h_A &= 150 \text{ ft} + (1.0 \text{ ft} + 23.4 \text{ ft}) \\ h_A &= 139.4 \text{ ft} \quad (140 \text{ ft}) \end{aligned}$$

The answer is (A).

Why Other Options Are Wrong

(B) This answer is incorrect because the height of the open tank is not equal to the TDH.

(C) This answer is incorrect because the 35 ft from the left side of the Bernoulli equation was not subtracted from the right side of the equation.

(D) This answer is incorrect because the 35 ft from the left side of the Bernoulli equation was added to, not subtracted from, the right side of the equation.

SOLUTION 8

In conventional welding engineering practice, the external loading in a fillet weld is carried as a pure shear stress in the throat area. The shear stress in one fillet weld is given by

$$\tau = \frac{F}{0.707hL}$$

The throat area is $0.707hL$ for each weld. Since there are two welds, one on the inside and one on the outside, the average shear stress is

$$\tau = \frac{F}{1.414hL}$$

The allowable shear stress, τ, is known, but the shear force, F, and length of the weld, L, are not. The length of the weld is the circumference of the center pipe.

$$L = \pi d = \pi(44 \text{ in}) = 138 \text{ in}$$

The force can be found by calculating the strength of the 38 bolts. Since the bolts are subjected to uniaxial stress, the maximum normal stress is

$$\sigma = \frac{F}{A}$$

The effective cross-sectional area for a ½ in UNF 20 bolt is 0.1599 in^2, and the tensile stress for an ASTM A354 grade BC bolt is 125 ksi.

The external load for each bolt is

$$\begin{aligned} F_{\text{bolt}} &= \sigma A \\ &= \left(125 \frac{\text{kip}}{\text{in}^2}\right)\left(1000 \frac{\text{lbf}}{\text{kip}}\right)(0.1599 \text{ in}^2) \\ &= 19{,}987.5 \text{ lbf} \end{aligned}$$

The total external load is

$$\begin{aligned} F_{\text{total}} &= F_{\text{bolt}}(\text{no. of bolts}) \\ &= (19{,}987.5 \text{ lbf})(38) \\ &= 759{,}525 \text{ lbf} \end{aligned}$$

The weld depth is found to be

$$\begin{aligned} h &= \frac{(\text{strength factor})F}{1.414L\tau} \\ &= \frac{(2)(759{,}525 \text{ lbf})}{(1.414)(138 \text{ in})\left(35 \frac{\text{kip}}{\text{in}^2}\right)\left(1000 \frac{\text{lbf}}{\text{kip}}\right)} \\ &= 0.222 \text{ in} \quad (0.250 \text{ in}) \end{aligned}$$

The answer is (A).

Why Other Options Are Wrong

(B) This incorrect answer results when, in the calculation of h, the weld area is not doubled.

(C) This incorrect answer results when, in calculating the force on the bolts, the nominal cross-sectional area is used instead of the actual cross-sectional area, and the weld area is not doubled.

(D) This incorrect answer results when, in calculating the force on the bolts, the nominal cross-sectional area is used instead of the actual cross-sectional area, the weld area is not doubled, and an incorrect maximum bolt tensile stress of 150,000 psi is used.

SOLUTION 9

The root-locus sketch of the open-loop transfer function $G(s)$ shown here indicates that the closed-loop system is stable for all values of K.

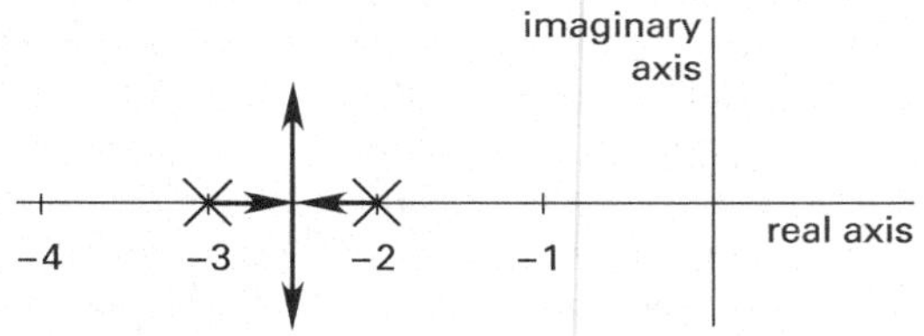

The answer is (D).

Why Other Options Are Wrong

(A) This incorrect answer is the value of the gain where the root locus breaks away from the negative real axis.

(B) This incorrect answer is the breakaway point on the negative real axis.

(C) This incorrect answer is part of the constant in the closed-loop characteristic equation.

Energy/Power System Apps.

SOLUTION 10

The counterflow configuration is as shown.

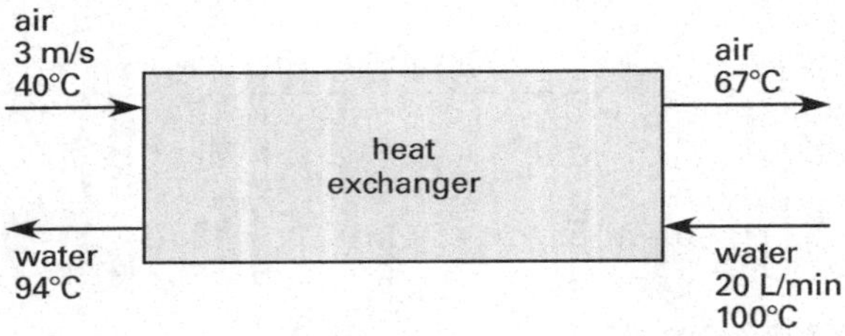

The heat transfer rate is

$$Q = U_0 A_0 F_c \Delta T_{\text{lm}}$$

F_c is the correction factor for a crossflow heat exchanger compared to a reference counterflow heat exchanger and ΔT_{lm} is the log mean temperature difference for a counterflow heat exchanger. The temperature differences are

$$\Delta T_A = 94°\text{C} - 40°\text{C} = 54°\text{C}$$
$$\Delta T_B = 100°\text{C} - 67°\text{C} = 33°\text{C}$$

The log mean temperature difference is

$$\Delta T_{\text{lm}} = \frac{\Delta T_A - \Delta T_B}{\ln \frac{\Delta T_A}{\Delta T_B}} = \frac{54°\text{C} - 33°\text{C}}{\ln \frac{54°\text{C}}{33°\text{C}}} = 42.6°\text{C}$$

The correction factor, F_c, is found using a correction chart for a single-pass, crossflow heat exchanger with both flows unmixed. The capacity ratio, R, and the efficiency, P, are

$$R = \frac{T_{\text{water,in}} - T_{\text{water,out}}}{T_{\text{air,out}} - T_{\text{air,in}}} = \frac{100°\text{C} - 94°\text{C}}{67°\text{C} - 40°\text{C}} = 0.22$$

$$P = \frac{T_{\text{air,out}} - T_{\text{air,in}}}{T_{\text{water,in}} - T_{\text{air,in}}} = \frac{67°\text{C} - 40°\text{C}}{100°\text{C} - 40°\text{C}} = 0.45$$

From an F_c correction chart, F_c is approximately equal to 1.0. The heat transfer rate is

$$Q = U_0 A_0 F_c \Delta T_{\text{lm}} = \left(62.9 \ \frac{\text{W}}{\text{m}^2\cdot°\text{C}}\right)(3.1 \ \text{m}^2)(1.0)(42.6°\text{C}) = 8307 \ \text{W} \quad (8.3 \ \text{kW})$$

The answer is (B).

Why Other Options Are Wrong

(A) This incorrect answer results when the area 0.37 m^2 is used.

(B) This incorrect answer results when the log mean temperature difference is calculated incorrectly.

(D) This incorrect answer results when incorrect values are used for the coefficient of heat transfer and the log mean temperature difference.

SOLUTION 11

The throttling area in a needle valve is a function of the seat opening diameter, needle lift, and cone half angle.

The answer is (C).

Why Other Options Are Wrong

(A) This answer is incorrect because throttling area is a function of needle lift.

(B) This answer is incorrect because throttling area is a function of seat opening diameter.

(D) This answer is incorrect because throttling area is a function of needle lift, and not a function of cone height.

SOLUTION 12

First convert the required flow of 23 in^3/sec to gallons per minute.

$$Q_{\text{req}} = \left(23 \ \frac{\text{in}^3}{\text{sec}}\right)\left(60 \ \frac{\text{sec}}{\text{min}}\right)\left(\frac{1 \ \text{gal}}{231 \ \text{in}^3}\right) = 5.97 \ \text{gal/min}$$

Begin by testing the 5 gpm valve because it is one of the middle answers. Calculating the constant KI_m using the manufacturer's equation and pressure drop value gives

$$KI_m = \frac{Q_R}{\sqrt{P_v}} = \frac{5 \ \frac{\text{gal}}{\text{min}}}{\sqrt{1000 \ \frac{\text{lbf}}{\text{in}^2}}} = 0.158 \ \text{gal-in/min-lbf}^{1/2}$$

Now the equation can be used again to determine what the flow rate through the 5 gpm valve would be for the actual pressure drop.

$$\begin{aligned} Q_{5\,\mathrm{gpm}} &= KI_m\sqrt{P_v} \\ &= \left(0.158\ \frac{\text{gal-in}}{\text{min-lbf}^{1/2}}\right)\sqrt{1500\ \frac{\text{lbf}}{\text{in}^2} - 1100\ \frac{\text{lbf}}{\text{in}^2}} \\ &= 3.16\ \text{gal/min} \end{aligned}$$

The flow rate that the 5 gpm value would give is less than the required flow rate, so this valve is not large enough. The quickest way to finish this problem from this point is to notice that Q is linear with respect to KI_m and that KI_m is linear with respect to Q_R. Thus, doubling Q_R to 10 gpm will double Q to 6.32 gpm, which satisfies the flow rate requirement.

Double-check the capacity of the 10 gpm value. The constant for this valve is

$$KI_m = \frac{10\ \frac{\text{gal}}{\text{min}}}{\sqrt{1000\ \frac{\text{lbf}}{\text{in}^2}}} = 0.316\ \text{gal-in/min-lbf}^{1/2}$$

The flow rate through this value under actual conditions is

$$\begin{aligned} Q_{10\ \mathrm{gpm}} &= \left(0.316\ \frac{\text{gal-in}}{\text{min-lbf}^{1/2}}\right) \\ &\quad \times\sqrt{1500\ \frac{\text{lbf}}{\text{in}^2} - 1100\ \frac{\text{lbf}}{\text{in}^2}} \\ &= 6.32\ \text{gal/min} \end{aligned}$$

This is greater than the required flow rate, so the 10 gpm valve will suffice. The 15 gpm valve would also provide a large enough flow rate, but it is probably more expensive than the 10 gpm valve. So the best choice is the 10 gpm valve.

The answer is (C).

Why Other Options Are Wrong

(A) This incorrect solution results from not converting seconds to minutes in the calculation of the required flow rate.

(B) This incorrect solution results from overlooking that the flow rate of 5 gpm quoted by the manufacturer is based on a pressure drop of 1000 psi. In this particular application the pressure drop is only 400 psi, so the actual flow rate through the valve would be significantly less than 5 gpm.

(D) This valve has a large enough flow rate, but it is not the least expensive valve that would work in the application.

SOLUTION 13

The thermal energy provided by the combustion of natural gas is given by the equation

$$\begin{aligned} Q_{\text{comb}} &= \eta \dot{m} h_{\text{HHV}} \\ &= (0.55)\left(65{,}000\ \frac{\text{lbm}}{\text{hr}}\right)\left(17{,}000\ \frac{\text{Btu}}{\text{lbm}}\right) \\ &= 607{,}750{,}000\ \text{Btu/hr} \end{aligned}$$

The amount of thermal energy needed to increase the temperature of the natural gas from 60°F to 980°F is

$$\begin{aligned} Q_{\text{temp}} &= \dot{m} c_p (T_2 - T_1) \\ &= \left(65{,}000\ \frac{\text{lbm}}{\text{hr}}\right)\left(0.692\ \frac{\text{Btu}}{\text{lbm-°F}}\right) \\ &\quad \times (980°\text{F} - 60°\text{F}) \\ &= 41{,}381{,}600\ \text{Btu/hr} \end{aligned}$$

The thermal energy available to heat the flue gas is

$$\begin{aligned} \Delta Q &= Q_{\text{comb}} - Q_{\text{temp}} \\ &= 607{,}750{,}000\ \frac{\text{Btu}}{\text{hr}} - 41{,}381{,}600\ \frac{\text{Btu}}{\text{hr}} \\ &= 566{,}368{,}400\ \text{Btu/hr} \end{aligned}$$

The mass flow rate of the flue gas after addition of the natural gas flow is

$$\begin{aligned} \dot{m} &= \dot{m}_{\text{flue}} + \dot{m}_{\text{feed}} \\ &= 3{,}180{,}000\ \frac{\text{lbm}}{\text{hr}} + 65{,}000\ \frac{\text{lbm}}{\text{hr}} \\ &= 3{,}245{,}000\ \text{lbm/hr} \end{aligned}$$

The amount of energy needed to increase the flue gas temperature is given by the equation

$$Q = mc_p(T_{f2} - T_{f1})$$

Solving for T_{f2},

$$\begin{aligned} T_{f2} &= \frac{Q}{\dot{m} c_p} + T_{f1} \\ &= \frac{566{,}368{,}400\ \frac{\text{Btu}}{\text{hr}}}{\left(3{,}245{,}000\ \frac{\text{lbm}}{\text{hr}}\right)\left(0.248\ \frac{\text{Btu}}{\text{hr-°F}}\right)} + 980°\text{F} \\ &= 1683°\text{F}\quad (1700°\text{F}) \end{aligned}$$

The answer is (B).

Energy/Power System Apps.

Why Other Options Are Wrong

(A) This incorrect answer results when the 980°F temperature increase is neglected.

(C) This incorrect answer results when the energy needed to heat the natural gas and the change in mass flow rate from the addition of the natural gas are neglected.

(D) This answer results when temperatures are mistakenly converted to Rankine.

SOLUTION 14

The compressor horsepower is

$$W_{\text{in}} = \dot{m}(h_2 - h_1)$$

The enthalpy at state 1, h_1, is given as 102.73 Btu/lbm.

The enthalpy at state 2, h_2, can be found from the superheat tables (approximately 119 Btu/lbm) or calculated from the COP equation.

$$\begin{aligned} \text{COP} &= \frac{h_1 - h_4}{h_2 - h_1} \\ h_2 &= \frac{h_1 - h_4}{\text{COP}} + h_1 \\ &= \frac{102.73 \ \frac{\text{Btu}}{\text{lbm}} - 41.787 \ \frac{\text{Btu}}{\text{lbm}}}{3.8} + 102.73 \ \frac{\text{Btu}}{\text{lbm}} \\ &= 118.8 \ \text{Btu/lbm} \end{aligned}$$

To find the mass flow rate of refrigerant, $\dot{m}$, rearrange the equation for the heat flow into the evaporator.

$$\begin{aligned} Q_{\text{in}} &= \dot{m}(h_1 - h_4) \\ \dot{m} &= \frac{Q_{\text{in}}}{h_1 - h_4} \\ &= \frac{(5 \ \text{tons})\left(12{,}000 \ \frac{\frac{\text{Btu}}{\text{hr}}}{\text{ton}}\right)}{102.73 \ \frac{\text{Btu}}{\text{lbm}} - 41.787 \ \frac{\text{Btu}}{\text{lbm}}} \\ &= 984.5 \ \text{lbm/hr} \end{aligned}$$

The compressor horsepower is

$$\begin{aligned} W_{\text{in}} &= \dot{m}(h_2 - h_1) \\ &= \frac{\left(984.5 \ \frac{\text{lbm}}{\text{hr}}\right)\left(118.8 \ \frac{\text{Btu}}{\text{lbm}} - 102.73 \ \frac{\text{Btu}}{\text{lbm}}\right)}{2545 \ \frac{\text{Btu}}{\text{hp-hr}}} \\ &= 6.216 \ \text{hp} \quad (6.2 \ \text{hp}) \end{aligned}$$

The answer is (B).

Why Other Options Are Wrong

(A) This incorrect answer results when the answer is calculated in kilowatts instead of horsepower.

(C) This incorrect answer results when h_4 is added to h_1 in the numerator of the COP equation instead of subtracted.

(D) This incorrect answer results when the value for h_4 is used in the final calculation instead of the value for h_1.

SOLUTION 15

When the torque is applied, only one cable in each pair of cables will be in tension, and the other cable will go slack. The torque resisted by the x-components of the forces in the cable is

$$T = 3F_x a$$

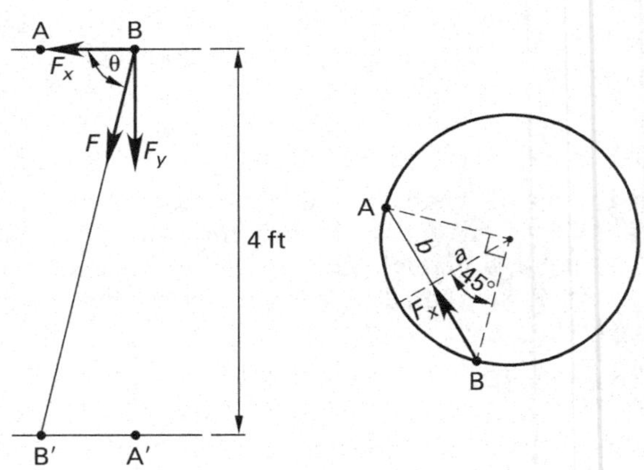

The perpendicular distance, a, from the center of the ring to F_x is

$$\begin{aligned} a &= \left((1 \ \text{ft})\left(12 \ \frac{\text{in}}{\text{ft}}\right)\right)\cos 45° \\ &= 8.485 \ \text{in} \end{aligned}$$

Use the torque equation to find the x-component of the force in the wire, F_x.

$$T = 3F_x a$$
$$F_x = \frac{T}{3a} = \frac{1800 \text{ in-lbf}}{(3)(8.485 \text{ in})} = 70.71 \text{ lbf}$$

To find the force in the wire, F, the angle θ is needed (see problem illustration). b is equal to a because they are the legs of a 45° right triangle, so

$$\theta = \arctan \frac{h}{2b} = \arctan \frac{(4 \text{ ft})\left(12 \frac{\text{in}}{\text{ft}}\right)}{(2)(8.485 \text{ in})} = 70.53°$$

Find the force F from its x-component.

$$F_x = F \cos\theta$$
$$F = \frac{F_x}{\cos\theta} = \frac{70.71 \text{ lbf}}{\cos 70.53°} = 212.1 \text{ lbf}$$

The normal stress in the cable is

$$\sigma = \frac{P}{A}$$

The cross-sectional area of the cable is

$$A = \frac{\pi d^2}{4} = \frac{\pi(0.0625 \text{ in})^2}{4} = 0.003068 \text{ in}^2$$

Because the cable is stranded, the actual cross-sectional area is reduced by a fill factor. The fill factor for 1 × 19 stranded cable is 0.76. The normal stress in the cable, then, is

$$\sigma = \frac{\left(\frac{212.1 \text{ lbf}}{1000 \frac{\text{lbf}}{\text{kip}}}\right)}{(0.003068 \text{ in}^2)(0.76)} = 90.96 \text{ kips/in}^2 \quad (91 \text{ ksi})$$

The answer is (D).

Why Other Options Are Wrong

(A) This incorrect answer results when an angle of 45° is used instead of 70.53°.

(B) This incorrect answer results when six cables are assumed to be in tension instead of three.

(C) This incorrect answer results when no fill factor is used.

SOLUTION 16

The modified Goodman fatigue failure criterion is

$$\frac{\sigma_a}{S_e} + \frac{\sigma_m}{S_{ut}} = \frac{1}{n}$$

For completely reversed loading, $\sigma_a = \sigma_{max}$. There are two possible critical points on the beam and they are at points B and C. At point B,

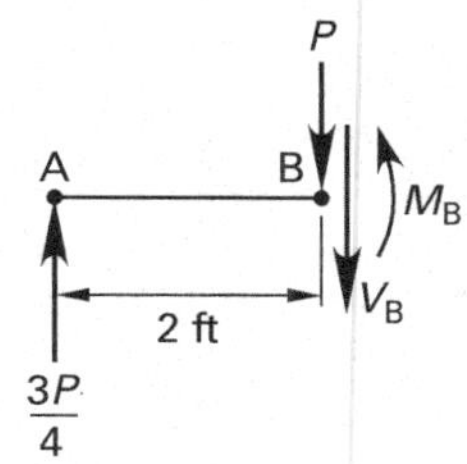

$$\sum M_{V_B} = 0$$
$$(24 \text{ in})\left(-\tfrac{3}{4}P\right) + M_B = 0$$
$$M_B = (18 \text{ in})P$$

There is no stress concentration at point B, so

$$\sigma_B = \frac{M_B c_B}{I_B} = \frac{M_B c_B}{\frac{td^3}{12}} = \frac{\big((18 \text{ in})P\big)(1.5 \text{ in})}{\frac{(2 \text{ in})(3 \text{ in})^3}{12}} = \frac{6P}{1 \text{ in}^2}$$

At point C,

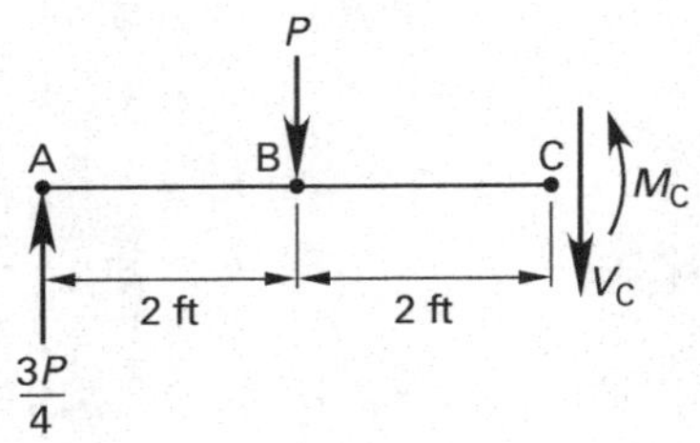

$$\sum M_{V_C} = 0$$

$$(48\text{ in})\left(-\tfrac{3}{4}P\right) + (24\text{ in})P + M_C = 0$$

$$M_C = (12\text{ in})P$$

Taking into account the stress concentration at point C, and using the notch sensitivity to reduce the theoretical stress concentration factor, K_t (see Shigley, *Mechanical Engineering Design*),

$$\sigma_{max} = K_f \sigma_C$$

$$K_f = 1 + q(K_t - 1)$$

To use the chart of theoretical stress concentration factors for a notched bar in bending from *Mechanical Engineering Design*, calculate r/d and w/d.

$$\frac{r}{d} = \frac{0.125\text{ in}}{2.75\text{ in}} = 0.045$$

$$\frac{w}{d} = \frac{3.0\text{ in}}{2.75\text{ in}} = 1.09$$

These values give a stress concentration factor, K_t, of approximately 2.4.

From the notch sensitivity chart for steel in *Mechanical Engineering Design*, using a radius of 0.125 in and an ultimate tensile strength for 1018 steel, S_{ut}, of 49.5 ksi, the notch sensitivity factor, q, is approximately 0.75. The stress concentration factor is therefore

$$K_f = 1 + q(K_t - 1) = 1 + (0.75)(2.4 - 1) = 2.0$$

The stress at point C is

$$\sigma_{max} = K_f\left(\frac{M_C c_C}{I_C}\right) = K_f\left(\frac{M_C\left(\frac{d_C}{2}\right)}{\frac{td^3}{12}}\right) = (2.0)\left(\frac{((12\text{ in})P)\left(\frac{2.75\text{ in}}{2}\right)}{\frac{(2\text{ in})(2.75\text{ in})^3}{12}}\right) = \frac{9.52P}{1\text{ in}^2}$$

Point C is the critical point. For completely reversed loading, the mean stress, σ_m, is equal to zero. For materials with S_{ut} less than 200 ksi, the endurance limit is

$$S_e' = 0.5S_{ut}$$

All the endurance limit modifying factors equal one, so the modified endurance limit is

$$S_e = S_e' = 0.5S_{ut} = (0.5)\left(49.5\ \frac{\text{kips}}{\text{in}^2}\right) = 24.75\text{ kips/in}^2$$

Use the modified Goodman equation and solve for P. n is the safety factor.

$$\frac{\sigma_a}{S_e} + \frac{\sigma_m}{S_{ut}} = \frac{9.52P}{S_e} + \frac{\sigma_m}{S_{ut}} = \frac{1}{n}$$

$$P = \left(\frac{S_e}{9.52}\right)\left(\frac{1}{n} - \frac{\sigma_m}{S_{ut}}\right) = \left(\frac{\left(24.75\ \frac{\text{kips}}{\text{in}^2}\right)\left(1000\ \frac{\text{lbf}}{\text{kip}}\right)}{9.52}\right) \times \left(\frac{1}{3} - \frac{0\ \frac{\text{kips}}{\text{in}^2}}{49.5\ \frac{\text{kips}}{\text{in}^2}}\right) = 866.6\text{ lbf} \quad (870\text{ lbf})$$

The answer is (B).

Why Other Options Are Wrong

(A) This incorrect answer results when K_t is used instead of K_f.

(C) This incorrect answer results when the stress at point B is used instead of the stress at point C.

(D) This incorrect answer results when the stress concentration and notch sensitivity at point C are ignored (that is, $K_f = 1$).

SOLUTION 17

The efficiency of the combined cycle is

$$\eta = \frac{w_{\text{net}}}{q_{\text{in}}}$$

The heat transfer into the system, q_{in}, is given as 340 Btu/lbm. Although the steam and gas mass flow rates are not given, their ratio is given as 0.131, so the net work done by the system per unit mass can be found.

$$\begin{aligned} w_{\text{net}} &= w_{\text{net,gas}} + \left(\frac{\dot{m}_{\text{steam}}}{\dot{m}_{\text{gas}}}\right) w_{\text{net,stream}} \\ &= 90\ \frac{\text{Btu}}{\text{lbm}} + (0.131)\left(570\ \frac{\text{Btu}}{\text{lbm}}\right) \\ &= 164.7\ \text{Btu/lbm} \end{aligned}$$

The efficiency of the combined cycle is therefore

$$\begin{aligned} \eta &= \frac{w_{\text{net}}}{q_{\text{in}}} \\ &= \frac{164.7\ \frac{\text{Btu}}{\text{lbm}}}{340\ \frac{\text{Btu}}{\text{lbm}}} \\ &= 0.4844 \quad (48\%) \end{aligned}$$

The answer is (C).

Why Other Options Are Wrong

(A) This incorrect answer results from multiplying together the individual efficiencies of the gas turbine cycle and steam cycle.

(B) This incorrect answer results when only the net work per unit mass of the steam cycle, adjusted by the mass flow rate ratio, is used in the final calculation instead of the net work done by the combined cycle.

(D) This incorrect answer results when the net work per unit mass of the steam cycle, adjusted by the mass flow rate ratio, is divided by the net work per unit mass of the gas turbine cycle to get the efficiency, instead of dividing by the heat transfer per unit mass.

SOLUTION 18

The thrust generated by the return stroke is

$$\begin{aligned} F &= \eta p(A_{\text{piston}} - A_{\text{rod}}) \\ &= \eta p\left(\frac{\pi}{4}\right)(d^2_{\text{piston}} - d^2_{\text{rod}}) \\ &= (0.85)\left(100\ \frac{\text{lbf}}{\text{in}^2}\right)\left(\frac{\pi}{4}\right)\left((3.0\ \text{in})^2 - (0.125\ \text{in})^2\right) \\ &= 599.8\ \text{lbf} \quad (600\ \text{lbf}) \end{aligned}$$

The answer is (C).

Why Other Options Are Wrong

(A) This incorrect answer results when the diameters are not squared in the calculation.

(B) This incorrect answer results when a pressure of 50 psi is inadvertently used instead of 100 psi.

(D) This incorrect answer results when an efficiency of 1.0 is used.